Central Nervous System Tissue Engineering

Current Considerations and Strategies

Synthesis Lectures on Tissue Engineering

Editor
Kyriacos A. Athanasiou and J. Kent Leach, *University of California, Davis*

The Synthesis Lectures on Tissue Engineering series will publish concise books on aspects of a field that holds so much promise for providing solutions to some of the most difficult problems of tissue repair, healing, and regeneration. The field of Tissue Engineering straddles biology, medicine, and engineering, and it is this multi-disciplinary nature that is bound to revolutionize treatments for a plethora of tissue and organ problems. Central to Tissue Engineering is the use of living cells with a variety of biochemical or biophysical stimuli to alter or maximize cellular functions and responses. However, in addition to its therapeutic potentials, this field is making significant strides in providing diagnostic tools.

Each book in the Series will be a self-contained treatise on one subject, authored by leading experts. Books will be approximately 65-125 pages. Topics will include 1) Tissue Engineering knowledge on particular tissues or organs (e.g., articular cartilage, liver, cardiovascular tissue), but also on 2) methodologies and protocols, as well as 3) the main actors in Tissue Engineering paradigms, such as cells, biomolecules, biomaterials, biomechanics, and engineering design. This Series is intended to be the first comprehensive series of books in this exciting area.

Central Nervous SystemTissue Engineering: Current Considerations and Strategies
Ashley E. Wilkinson, Aleesha M. McCormick, and Nic D. Leipzig
2011

Biologic Foundations for Skeletal Tissue Engineering
Ericka M. Bueno and Julie Glowacki
2011

Regenerative Dentistry
Mona K. Marei
2010

Cells and Biomaterials for Intervertebral Disc Regeneration
Sibylle Grad, Mauro Alini, David Eglin, Daisuke Sakai, Joji Mochida, Sunil Mahor, Estelle Collin, Biraja Dash, and Abhay Pandit
2010

Fundamental Biomechanics in Bone Tissue Engineering
X. Wang, J.S. Nyman, X. Dong, H. Leng, and M. Reyes
2010

Articular Cartilage Tissue Engineering
Kyriacos A. Athanasiou, Eric M. Darling, and Jerry C. Hu
2009

Tissue Engineering of Temporomandibular Joint Cartilage
Kyriacos A. Athanasiou, Alejandro J. Almarza, Michael S. Detamore, and Kerem N. Kalpakci
2009

Engineering the Knee Meniscus
Kyriacos A. Athanasiou and Johannah Sanchez-Adams
2009

Central Nervous SystemTissue Engineering: Current Considerations and Strategies

Ashley E. Wilkinson, Aleesha M. McCormick, and Nic D. Leipzig

ISBN: 978-3-031-01454-3   paperback
ISBN: 978-3-031-02582-2   ebook

DOI 10.1007/978-3-031-02582-2

A Publication in the Springer Nature series
SYNTHESIS LECTURES ON ADVANCES IN AUTOMOTIVE TECHNOLOGY

Lecture #8
Series Editor: Kyriacos A. Athanasiou and J. Kent Leach, *University of California, Davis*
Series ISSN
Synthesis Lectures on Tissue Engineering
Print 1944-0316   Electronic 1944-0308

Central Nervous System Tissue Engineering

Current Considerations and Strategies

Ashley E. Wilkinson, Aleesha M. McCormick, and Nic D. Leipzig
The University of Akron

SYNTHESIS LECTURES ON TISSUE ENGINEERING #8

ABSTRACT

Combating neural degeneration from injury or disease is extremely difficult in the brain and spinal cord, i.e. central nervous system (CNS). Unlike the peripheral nerves, CNS neurons are bombarded by physical and chemical restrictions that prevent proper healing and restoration of function. The CNS is vital to bodily function, and loss of any part of it can severely and permanently alter a person's quality of life. Tissue engineering could offer much needed solutions to regenerate or replace damaged CNS tissue. This review will discuss current CNS tissue engineering approaches integrating scaffolds, cells and stimulation techniques. Hydrogels are commonly used CNS tissue engineering scaffolds to stimulate and enhance regeneration, but fiber meshes and other porous structures show specific utility depending on application. CNS relevant cell sources have focused on implantation of exogenous cells or stimulation of endogenous populations. Somatic cells of the CNS are rarely utilized for tissue engineering; however, glial cells of the peripheral nervous system (PNS) may be used to myelinate and protect spinal cord damage. Pluripotent and multipotent stem cells offer alternative cell sources due to continuing advancements in identification and differentiation of these cells. Finally, physical, chemical, and electrical guidance cues are extremely important to neural cells, serving important roles in development and adulthood. These guidance cues are being integrated into tissue engineering approaches. Of particular interest is the inclusion of cues to guide stem cells to differentiate into CNS cell types, as well to guide neuron targeting. This review should provide the reader with a broad understanding of CNS tissue engineering challenges and tactics, with the goal of fostering the future development of biologically inspired designs.

KEYWORDS

central nervous system, tissue engineering, spine regeneration, spinal cord injury, brain injury, neurodegenerative disease, nerve guidance, neural stem cells, nerve scaffold, neurotrophic factors

Contents

CHAPTER 1

Introduction

The brain and spinal cord compose the Central Nervous System (CNS), which is the control center of the body. Inputs from muscles, involuntary organs, and senses travel through the nerves of the Peripheral Nervous System (PNS) into the CNS where they are interpreted. Signals may travel within the brain to separate functional areas. Instructions are then sent outward again for voluntary movement and involuntary regulation to complete the endless loop of the nervous system circuitry. The CNS is critical to function of the entire body, which is why incurred injury and disease cripple one's quality of life. In the United States alone approximately 265,000 people are estimated to have spinal cord injuries (SCIs), with over 10,000 new injuries occurring each year [1]. Patients with SCI experience decreased lifespan in addition to life-costs from one to four million dollars, depending on the extent of injury, which is especially disturbing considering the fact that the average age of a spinal cord injured person is 31 [1]. Moreover, traumatic brain injuries (TBIs) occur to over 1.7 million people each year [2]. The devastating physical and psychological effects of CNS damage are felt by both patients and their families. Of SCI individuals experiencing paraplegia or tetraplegia, less than 1% achieve full neurological recovery post treatment [1]. Solutions to recover neurological function are desperately needed for all CNS injuries.

Tissue engineering (TE) in the CNS is extremely difficult because of the intrinsic restrictions and complexity of native CNS tissue. Generally, multi-component approaches are used in attempt to restore natural function to the brain or spinal cord. First and foremost, an understanding of tissue formation and function as well as tissue responses to damage is needed in order to formulate treatments to correct injury and disease in the CNS. Knowledge of native tissue and pathological development will foster improvement of strategies for overcoming damage to the CNS. For the most severe CNS disorders, a complex TE construct involving multiple cues is most likely needed to combat the physical and chemical obstacles of the CNS; the general building blocks of these constructs are physical scaffolding from biomaterials, endogenous or exogenous cells, and stimulatory cues from chemical, mechanical and electrical signals within the construct or on its surface (Fig. 1.1). Within this review, scaffold formation techniques and common biomaterials in CNS TE will be discussed followed by potential cell sources. Subsequent sections will discuss the myriads of stimulatory and guidance techniques currently being employed in CNS strategies. The hope of this review is to give the reader the basic tools for designing or understanding strategies aimed at regenerating the CNS and also for exposure to current approaches. PNS regenerative strategies are often discussed to augment basic understanding and many of these techniques do translate to the CNS. In depth review articles and books are suggested throughout for further reading on particular topics.

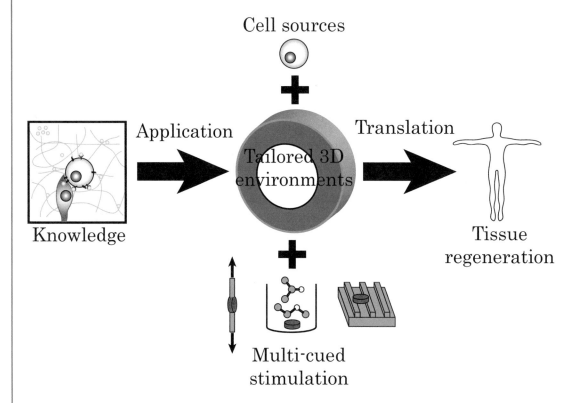

Figure 1.1: General tissue engineering strategy: knowledge of native cell environment is used to combine cell sources with mechanical, chemical, and electrical cues into a tailored tissue engineering construct. This assembly is grown under proper stimulation for translation to host tissue for regeneration.

Anatomy of the CNS and Progression of Neurological Damage

2.1 ANATOMY AND PHYSIOLOGY OF THE CNS

2.1.1 GROSS ANATOMY

Neurons throughout the body are organized into three main structures: ganglia, nuclei, and lamina [3]. Outside of the CNS, neuronal bodies are grouped together into ganglia; an example of this formation is the dorsal root ganglia (DRG) of sensory cell bodies that form just outside of the spinal column in the PNS. Within the brain, neuronal bodies with a common function are grouped together into nuclei. The bulk of the brain is organized into a layered cortex, including most of the cerebrum and cerebellum. The cerebrum, diencephalon, cerebellum, and brain stem make up the parts of the brain and are all housed within the skull (Fig. 2.1). The cerebral cortex is extremely complex and has a number of functions including, but not limited to, systemic sensory and motor control, speech, recognition and understanding [4]. The spinal cord runs inferior to the brain stem in a columnar form, protected by the vertebrae of the spine. The inner core of the spinal column, with a butterfly like shape, contains the gray matter while the surrounding axons are white matter. Dorsal horn (sensory), ventral horn (motor), intermediate zone and commissural region comprises the gray matter [3]. White matter is made up of the anterior, posterior and lateral columns [3]. Gray matter is largely unmyelinated while myelination provides white matter with its name and color. Nerve fibers enter and exit the spine at each vertebra though holes called Foramen, allowing information to pass to and from the PNS. There are four main groups of spinal nerves that exit at different levels of the spinal cord. Named in descending order down the vertebral column, these are cervical (neck), thoracic (upper back), lumbar (lower back) and sacral (base) nerves. While the PNS is made up of groups of axons termed nerves, in the CNS axons run in groups called tracts that are bound together by the processes of astrocytes, often called 'end-feet' [3]. Descending (efferent) pathways include the pyramidal and extrapyramidal tracts, and ascending (afferent) pathways include the spinothalamic and spinocerebellar tracts as well as the gracile and cuneate fasciculi. Proper bodily function depends on these paths to transmit information between the brain and periphery.

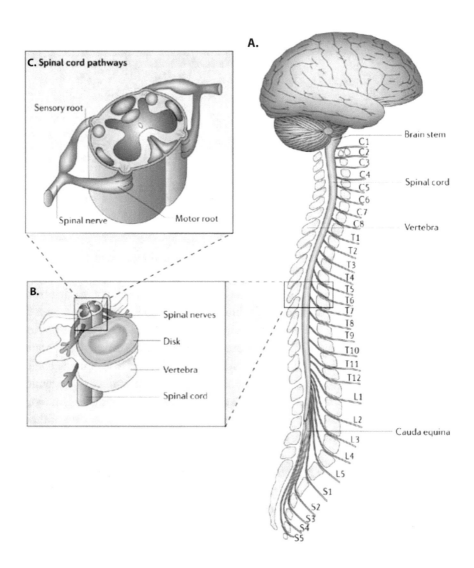

Figure 2.1: (A) Illustration of the CNS displaying the brain and spinal cord. (B) Cross-section of the spinal column showing the spinal cord protected within the vertebrae. (C) The spinal cord is segregated with white matter surrounding gray matter. Spinal roots exit on the ventral side and enter on the dorsal side to and from the PNS. Figure reprinted from [493].

2.1.2   EXTRACELLULAR MATRIX

The extracellular matrix (ECM) is an important component of the CNS, and it accounts for around 20% of the adult brain [5]. More thorough reviews of CNS ECM are described elsewhere; only a few of the common ECM molecules are discussed here [5, 6]. The brain and spinal cord are primarily made up of proteins and proteoglycans - macromolecules with a protein core and glycosaminoglycan (GAG) side chains. GAGs are linear, negatively charged polymers of repeating disaccharide units also present in the extracellular space that help to properly hydrate tissues. Not only does the ECM provide the natural scaffolding for tissue, but it plays an active role in the regulation of diffusion of soluble proteins and in localizing membrane proteins to functional domains. Collagens and laminins are the primary ECM proteins of the CNS, which mainly make-up the basal lamina, and contain specific amino acid sites that interact with cell receptors [5, 7]. Several types of collagens are found in the brain and spinal cord (I, II, IV, XVII, XIX); however, they are not as abundant in the CNS as they are in most other tissues [8]. Integrins are the receptors that cells use to interact with the ECM. These transmembrane glycoproteins are made up of an alpha and a beta subunit that complex to activate a host of signaling pathways within the cell. Integrins provide a link from the ECM to the cytoskeleton. In the CNS, $\beta1$ integrins are most relevant, binding to ligand sites on laminins, and are critical for proper neuronal migration [6, 7]. A major GAG found in the CNS and important constituent of the brain ECM is hyaluronic acid (HA) [5]. HA is implicated in many cellular functions, including regulating the diffusion of synaptic elements. Specifically, HA rich ECM in the synaptic region of the brain serves to restrict the escape of neurotransmitters and provides a physical barrier preventing the diffusion of the post-synaptic α-amino-3-hydroxy-5-methyl-4-isoxazolepropionic acid (AMPA) receptor to other areas of the plasma membrane [5]. AMPA receptors are cell ion channels that allow current to pass when activated by bound glutamate and are involved in fast synaptic transmission in the CNS [9]. In this way, HA contributes to maintaining proper synaptic signaling. Chondroitin sulfate proteoglycans (CSPGs) are the most common proteoglycans in the CNS, and include aggrecan, brevican, neurocan, and versican [10]. These CSPGs are termed lecticans due to their lectin-like domain, or a sugar binding domain [11]. Along with heparan sulfate proteoglycans (HSPGs), CSPGs are known to inhibit axon regeneration, but have been implicated in growth factor retention and presentation in healthy CNS tissue [5, 6].

2.1.3   NEURONS

Neurons are the fundamental units of the nervous system that process and transmit information by chemical and electrical signaling. Neurons are regionalized specifically to carry out signaling and can either be efferent, sending information away from the brain; afferent, sending information toward the brain; or interneurons, sending information between functional groups of neurons. The dendrites and soma compile and interpret cues from other cells and the surrounding environment (Fig. 2.2A). The soma serves as the trophic center of the neuron, regulating and producing proteins to be sent to various parts of the neuron [3]. Located here are most of the organelles common to all eukaryotic cells, including the nucleus and endoplasmic reticulum. Most cytoplasmic components are made

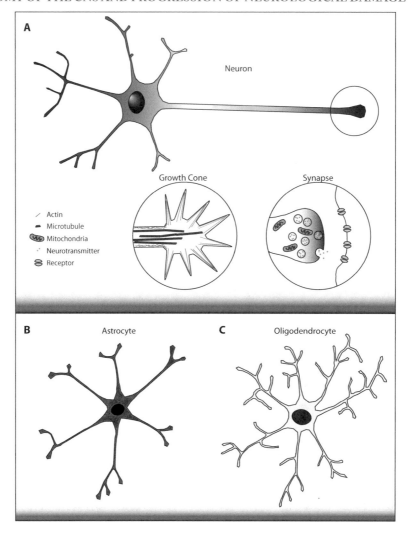

Figure 2.2: *Continued.* Major cell types of the CNS. (A) Different neurons have varied dendrite and axon conformations depending on their specific roles. The typical neuron has a large number of dendrites to gather information and a long axon to transmit the signal to its target. Insets show the terminus of the axon: Left: the growth cone is present during development and regeneration. Microtubules make up the core of the growth cone, while the peripheral lamellipodia and filopodia are comprised of depolymerized and F-actin. Right: once the axon finds its target it forms a synapse full of neurotransmitter vesicles and mitochondria. The neurotransmitters affect target receptors across the synaptic cleft. Illustration of an astrocyte in red (B) and oligodendrocyte in blue (C). Morphologically each cell follows its name; astrocytes are generally star shaped while oligodendrocytes have many branches, allowing them to myelinate more than one neuron at a time.

in the soma and transported to the processes directly or in vesicles. Dendrites sprouting from the soma mainly function to provide space for synapse, or connection to other neurons, and interpret the summation of excitatory and inhibitory signals from other neurons or extracellular space. They take on different formations depending on the type of neuron, but dendrites are thicker than axons and generally have many branches, sometimes even containing protruding spines to enhance synaptic reception area [4]. Although synapses may form on the soma or axon, the majority of synapses are located in the dendrites of a neuron [3].

The axons are the fiber-optic pipeline of the neuron, passing information at high speeds. Electrical signals, or action potentials, travel away from the soma down the axon in an all-or-nothing manner. The cytoskeleton of the axon is made up of intermediate filaments aligned along the axon and some microtubules constructed from tubulin, although not as many as are found in the dendritic processes. Tubulin is a polymer of repeating α and β subunits whose stability is promoted by microtubule-associated proteins (MAPs) and specific nucleotides [4]. Actin microfilaments are found throughout the axon, but are particularly important at the terminal end of the axon, or the growth cone, and will be discussed in greater detail later. The axon can be shorter, as in a satellite neuron, or very long to carry signals great distances. For example, adult human spinal cord axons can reach a length of several feet [12]. Whatever the length, vesicles and proteins need to be transported quickly along the axon to and from the soma. This is accomplished by motor proteins that use ATP hydrolysis to travel along microtubules; kinesin is responsible for soma to terminus (anterograde transport), and dynein carries organelles or vesicles from the axon to the soma (retrograde transport) [4]. Axon transport is extremely important for delivering neurotrophins and neurotransmitters to the terminus, as well as bringing proteins back to the soma for reprocessing. Depending on the molecule or vesicle being relocated, rate of transport in the axon can range from around 1 to 300 mm/d [3].

Axons have developed an evolved way of traveling to their target cell; the terminus of the axon, or the growth cone, interprets growth and directional signaling molecules and guides the axon along its path. Growth cones contain microtubules in their core and actin in their periphery (Fig. 2.2A). Long filaments of actin (F-actin) are found in the spikes protruding from the growth cone (filopodia) and disassembled actin is found in the web like lamellipodia that are just proximal to the filopodia [4, 13]. Actin is extremely important to the dynamic movement of the growth cone. Permissive substrates facilitate focal adhesion attachments of the growth cone to the substrate and stabilization of F-actin keeps the filopodium from being retracted. The lamellipodia follows in tow, advancing the growth cone [13, 14]. New cytoskeleton and membrane must be added to the axon as the growth cone hastens forth; constant transport of membrane components to the terminus as well as protein synthesis within the growth cone itself allows significant advancement even if the cell machinery is distant [4, 15].

Once the growth cone has arrived at its innervating target, a synapse is formed, either chemical or electrical. The chemical synapse includes the pre-synaptic terminal of the axon, the post-synaptic area of the cell being acted upon, and the very small gap (30-40 nm) between the two termed the synaptic cleft [3]. The pre-synaptic area of the neuron is filled with mitochondria and synaptic vesicles containing neurotransmitters (Fig. 2.2A). When an action potential is generated along the axon it travels all the way to the synapse, leading to an increase in calcium levels that causes neurotransmitters to be exocytosed into the synaptic cleft. These neurotransmitters travel to receptors in the post-synaptic region, diffuse into surrounding tissue, or are endocytosed by nearby astrocytes [3–5]. Receptors activated by neurotransmitters change the permeability of the membrane to specific ions, which will lead to either a depolarization (excitatory response) or hyperpolarization (inhibitory response) of the target cell. Synapses in the dendritic region tend to be excitatory, while those on the soma are usually inhibitory [3]. In electrical synapses, neurons are connected by gap junctions that allow ions to pass directly from one cell to the next without mediation by chemical transmitters. In contrast to chemical synapses, electrical signaling has no delay, but is only excitatory, is not amplified, and may be bidirectional [4].

2.1.4   GLIA AND SUPPORTIVE TISSUE

The main role of astrocytes (Fig. 2.2B) in the CNS is supporting neuronal function by creating and protecting the neuronal microenvironment. There are over one hundred times more astrocytes in the CNS than neurons, and they maintain and mimic neurons directly by absorbing and releasing neurotransmitters into the synapse [3]. Astrocytes protect the CNS by forming the outer glial layer of the brain and enwrapping vasculature to create the blood-brain barrier that is infamous for its extreme selectivity [3]. The structure of the CNS is largely due to the action of astrocytes; they form the outer and inner glial membrane and isolate axons throughout the brain and spinal cord. In the event of injury, astrocytes proliferate and become a glial scar, which is a major blockade for neural TE and will be discussed in greater detail later this chapter.

Oligodendrocytes are the myelin forming cells of the CNS, serving to surround axons in an electron-dense myelin sheath. Processes of oligodendrocytes wrap around a neuron's axons many times, squeezing out most of the cytoplasm. Layers of lipid rich plasma membrane are left, tightly encircling and insulating the axon [4]. Oligodendrocytes have many processes (Fig. 2.2C) and a single oligodendrocyte can myelinate 30-60 axons at once. Myelin is an electrical insulator, whose main purpose is to increase the speed of electrical conduction through the axon while preventing signal loss. Electrical current cannot conduct through the myelinated portions of the axonal membrane, it only occurs at small gaps between myelin, termed nodes of Ranvier, which are several micrometers in length [3, 4]. Not all axons are myelinated, but the ones that are have much faster signal transmission times due to the saltatory conduction of current "jumping" from node to node along the axon. Injury to oligodendrocytes, and subsequent demyelination of axons, has been shown to lead to nervous system degeneration [16, 17]. Similarly, demyelination of axons is the causative factor for the symptoms of multiple sclerosis (MS).

Microglia are the immune cells of the CNS. Microglia originate from monocytes that have been trapped in the CNS during development and evolve into a less active state, or resting state [3]. At any sign of injury or disease these resting microglia proliferate and may become active, expressing class I major histocompatibility complex (MHC). Due to their reactivity, microglia are used in research to gauge the extent of an insult to the brain or spinal cord by detecting the amount of activated microglia in the area [18–20].

Ependymal cells serve to line the ventricles and central canal of the CNS. These cells, in conjunction with blood vessels in the brain, secrete cerebrospinal fluid (CSF) [3]. Cilia on the surface of ependymal cells circulate CSF in the ventricles. Recently, they have been found to possess plasticity, and have the ability to differentiate into glia of the CNS in response to specific stimuli [21–23].

2.2 LOSS OF NEURAL FUNCTION

Injury response and subsequent nerve regeneration is very different in the CNS and PNS, resulting in contrasting outcomes. Due to differences in the two healing environments, PNS axons are able to reinnervate their targets while CNS axons are inhibited by physical and chemical blockades.

2.2.1 MODEL SYSTEMS AND FUNCTIONAL RECOVERY EVALUATION

Throughout the course of this review, different models of evaluation for TE strategies will be discussed. For understanding the implications of particular situations, a basic knowledge of injury models and evaluation techniques is necessary. Simulating injury in the brain and spinal cord has been standardized to some degree to enable global comparisons of different treatments across research labs. Cut or crush injuries are generally discussed for the spinal cord. Cut injuries are mimicked in the laboratory by surgical removal of all or a columnar portion of the spinal cord, termed complete transection or hemisection, for half of the spinal cord. To simulate a crush injury, the spinal cord is usually exposed and an electromagnetic or weight drop device is used to contuse the tissue. Animal disease models also exist where a key feature of a disease (e.g., demyelination) is simulated genetically for comparison of therapeutic methods.

Once an injury or disease model is created for study, specific methods for gauging deterioration or recovery are used to gather results. "Functional recovery" is a very subjective term and is applied mainly to *in vivo* work but sometimes it can be applied to *in vitro* models as well. *Ex vivo* cell and tissue are analyzed in a number of ways including examination of the cell and tissue anatomy (through histology, immunohistochemistry (IHC), and microscopy) as well as particular DNA, RNA, protein, or receptor expression (via polymerase chain reaction (PCR), microarrays, enzyme-linked immunosorbent assay (ELISA) and patch clamping). In animal models, functional improvement is usually estimated using behavioral observations and a rating score. For motor function, exercise tests and open field walking tests are often used. A popular and systematically defined scoring system is the Basso, Beattie and Bresnahan (BBB) locomotor rating scale [24]. The designers of this scoring system observed rats walking in an open field and assigned points for movements in paw joints, plantar steps, coordination, and limb alignment. In addition, exercise tests are sometimes used to assess motor recovery including swimming, rotor clinging, and narrowing track tests [25]. For sensory evaluation, reflex tests are used; reaction time to a toe pinch or heat application correlate to scoring of the animal's sensory recovery.

2.2.2   AXON RETRACTION AND DEGENERATION

There are two modes through which axons remodel in the body, retraction (small scale pruning of axons and dendrites) and degeneration (large scale elimination of large portion of the primary axon or collateral branches). Retraction and Wallerian degeneration will be covered as well as possible mechanisms for these types of neuronal preservation models. For a more in depth understanding of these occurrences please see [16].

Axon Retraction

Retraction typically concerns small scale maintenance of the nervous system where multiple target innervations are eliminated by local pruning of axonal and dendritic branches. Two primary examples of this phenomenon occur at the primary visual cortex and the neuromuscular junction (NMJ), where a motor neuron axon synapses with muscle fiber motor end plate (Fig. 2.3). As an adult, NMJs are innervated with a single axon; however, during development multiple innervations take place at the same NMJ site. Each axon contends for this synapse and the competition leads to retraction of some axons due to the strengthening of others; finally, a single axon takes its place and vacant, noninnervated post-synaptic sites lose their receptors [26]. Retraction also leads to the highly ordered organization of the mammalian visual system. Retinal ganglion cells (RGCs) travel to the visual information relay center of the brain, the lateral geniculate nucleus (LGN), and neurons are further projected to the primary visual cortex, located in the occipital lobe of the cortex. Higher order neurons are then segregated into specific and complex patterns in this region allowing each eye to receive inputs for visual depth perception. During visual development these neurons overlap and mature connections are made through the pruning and segregation of axonal arbors. These adult, stable connections occur through synaptic plasticity proposed by Hebb's postulate where pre-synaptic efficiency results from the continual stimulation of post-synaptic cells [27].

Mechanisms of Axon Retraction

The mechanisms of retraction are currently poorly understood; as such, the basic definition of retraction is a change in cell shape, manipulation of the cytoskeleton, signaling pathways, or environmental cues. As described above, cytoskeletal components such as microtubules and actin are involved in growth cone migration. Axon retraction is a dynamic process involving interaction between these factors. Retraction of neurons results when microtubule polymerization is inhibited; whereas, retraction does not occur with blocked reduction in ATP microtubule assembly [28]. Inhibition of dynein on intact microtubules can lead to axon retraction; however, this does not occur when microfilaments are depleted [29]. Therefore, during retraction it is believed that motor molecules counterbalance changes in microfilaments. Cytoskeletal regulation and changes result from intercellular signaling cascades. Inhibition of rho-associated protein kinase (ROCK) blocks Ras homolog gene family, member A (RhoA) downstream signaling activation on axonal and dendritic retraction in hippocampal neurons [30]. Inhibitory guidance molecules (discussed in more detail in Chapter 5) located in the extracellular environment initiate RhoA signaling cascades, and are therefore thought

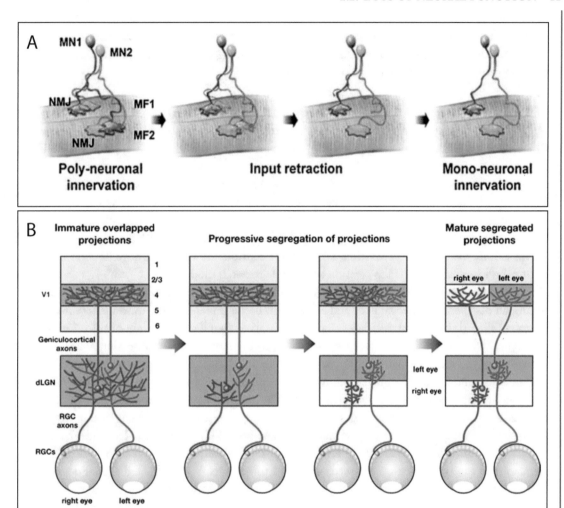

Figure 2.3: Two common axonal retraction examples during development. (A) Input signals from a motor neuron (MN) to a muscle fiber (MF) creates a neuromuscular junction (NMJ). During development, multiple synapses are formed (left). As MNs undergo synaptic plasticity, weaker neurons retract as a single neuron is victorious, resulting in a mature, adult NMJ (right). (B) Retinal ganglion cells (RGCs) are separated into specific layers of the dorsal lateral geniculate nucleus (dLGN). Geniculocortical axons are projected from this region and segregated into eye-specific columns in layer 4 of the primary visual cortex (V1). These synapses mature (from left to right) via competition from overlapping neurons. Small-scale elimination and retraction cause these overlapping monocular inputs to become isolated into specific regions of the dLGN and V1, producing a mature visual system. Images reprinted from [16].

to be key players in axon retraction [31, 32]. Even though many of the causative initiators of axon retraction remain unclear, it plays a significant role in neuronal development and the establishment of functional connections. Uncovering and fully understanding these mechanisms more clearly could result in better models and functional recovery outcomes.

Wallerian Degeneration and Slow Wallerian Degeneration

For Wallerian degeneration following injury, a latent period precedes rapid progression into active cytoskeletal breakdown, membrane blebbing, and axon fragmentation (Fig. 2.4) [16, 33–36]. Upon injury, degeneration occurs in 3-4 d; however, in lower vertebrates and invertebrates the axon segment can last for a period of more than ten times longer before degeneration proceeds [37]. Although Wallerian degeneration occurs after a cut or crush injury, it is also similar to axon degeneration observed in the later stages of some neurodegenerative diseases [16, 33]. Study of this response is invaluable for understanding axonal degeneration due to this commonality. PNS, CNS, and explanted nerves exhibit Wallerian degeneration, allowing for controllable Wallerian degeneration initiation *in vitro* and *in vivo* and thus the creation of valuable research models.

The discovery of the Wld^S mutant mouse by Lunn *et al.* in 1850 accelerated the study of Wallerian degeneration mechanisms and allowed for better understanding of some nervous system diseases [35, 36]. Lunn and associates originally observed that a special strain of mice was able to propagate action potentials for over two weeks following sciatic nerve transection, whereas, after injury, wild-type mice only carry action potentials for 1.5 d [38]. This delay was correlated to axon degeneration and facilitated further study of physical and molecular events that occur following nerve injury. Subsequent studies revealed that Wld^S axons degrade in a more gradual atrophic process whereas wild-type axon degeneration is self-regulated by pre-existing machinery in the axon, similar in manner but not in mechanism to apoptosis [16, 34, 39, 40]. The wide-held belief is that Wld^S mice have the ability to either defer or completely suppress Wallerian degeneration since injured axons die by different mechanisms in these mutants. Obstructed axonal degradation of Wld^S mice is dependent and a sole property of neurons themselves, requiring zero assistance from the neural system or glia of the mutant strain [41, 42]. Found mainly in the nucleus, Wld^S is a fusion protein containing nicotinamide mononucleotide adenylyltransferase 1 (Nmnat1), a specific segment of ubiquitin conjugation factor E4 B (Ube4b), and a unique 18 amino acid sequence joining the two [35, 43]. The question remains as to whether Wld^S protein acts from within the nucleus by recruiting and directing other pathways or if low concentrations within the axon are sufficient to interfere with degeneration. There is controversy over whether slowing degeneration is beneficial to injury recovery. Several studies have shown that in cases where degeneration was slowed using Wld^S mice; regeneration was also delayed and often muted after axotomy [44, 45]. While attempting to compare developmental pruning to axonal injury, Martin *et al.* found that regenerating axons in developing zebrafish avoided persistent fragments from induced axotomy [46]. The group speculated that slow clearance of degenerating axons may be detrimental to innervation.

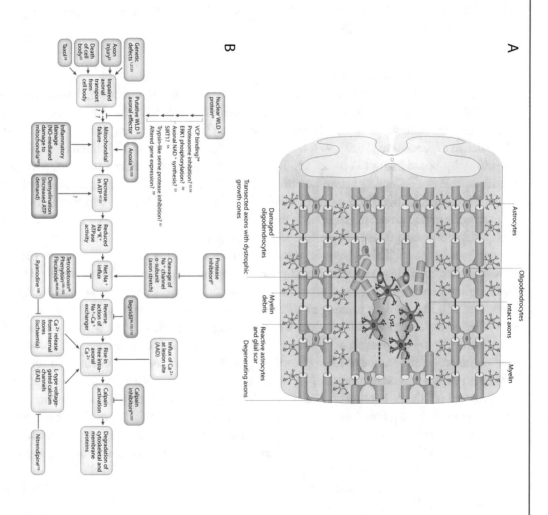

Figure 2.4: *Continues.*

Figure 2.4: *Continued.* (A) CNS injury response is extremely restrictive. Damage incurred in the spinal cord is followed by fragmentation of axons and myelin, with limited clearance. Reactive astrocytes form a glial scar, preventing the axons from resynapsis. (B) Axon degeneration pathways showing the scope and complexity of Wallerian degeneration. AAD, acute axon degeneration; EAE, experimental autoimmune encephalomyelitis; ERK-1, extracellular signal regulated kinase 1; NO, nitric oxide; SIRT 1, silent information regulator; VCP, valosin-containing protein; WLDS, slow Wallerian degeneration protein. Images reprinted from [57] and [33].

Mechanisms of Wallerian Degeneration
The complete molecular pathway for Wallerian degeneration is still unknown, despite the recent identification of several key players within the axon that are associated with the process (Fig. 2.4B). Axonal transport failure and microtubule dissociation are believed to be one of the earliest initiators of Wallerian degeneration [16, 33, 47]. Once the axon is lesioned, transport is limited or completely cut-off. A lag phase follows in which the distal axon is separated from the proximal portion, yet is still capable of conducting action potentials. Once the latent period is over, a catastrophic and active breakdown of the distal segment takes place via innate machinery within the axon. One culprit implicated in axonal degradation is the ubiquitin-proteasome system (UPS) [16, 33, 46, 47]. Ubiquitin acts to tag proteins for destruction via proteasomes, enzymes that denature proteins by peptide bond scission. This system is present in most cells and is important to many biological processes. Recent studies have shown that pharmacological and genetic inhibition of the UPS can significantly increase the lag time after axotomy, but only when administered before a lesion is made [46, 47]. The need for priming of UPS inhibition suggests its involvement in the early events of Wallerian degeneration. Additionally, a rise in intracellular calcium ion (Ca^{2+}) concentration is necessary for the progression of Wallerian degeneration and subsequent regeneration. Increased Ca^{2+} promotes cyclic adenosine monophosphate (cAMP) activity and neurofilament breakdown by calpain, a calcium dependent protease [47, 48]. Axon degeneration is often compared to apoptosis, or programmed cell death, because of similarities in the way each works on the cell including targeted ubiquitination [33–35]. Since axon degradation is performed by activation of similar cell proteins to apoptosis, studies have attempted to uncover similar pathways for the two processes. So far, research has revealed that the mechanisms of each are independent; moreover, when distal portions of injured axons were subject to nerve growth factor (NGF) deprivation, UPS inhibition resulted in delayed axon degeneration and only inhibition of apoptosis saved both the soma and axon of deprived neurons [47].

2.2.3   NEURODEGENERATIVE DISEASES
Wallerian-like degeneration has been observed in many neurodegenerative diseases, including Charcot-Marie-Tooth, MS, amyotrophic lateral sclerosis (ALS), Alzheimer's, Parkinson's, Huntington's, and prion diseases, such that axonal transport is disrupted without transection or crush

injury [16, 33–35, 49, 50]. The base cause of these diseases often occurs because of improper or missing axonal transport protein function, frequently implicating either microtubules or antero and retrograde motor proteins. Proteins and organelles become trapped in varicosities, or the minor axonal swellings, leading them to become spheroid formations (major swelling) [33]. Aberrant swelling and spheroid formation within the axon from loss of transport has been linked to initiation of Wallerian degeneration in disease models [33, 51, 52]. After spheroid formation, the axon degrades in a strikingly similar fashion to traditional Wallerian degeneration [51, 52]. Much of the focus of therapeutic Wallerian degeneration in neurodegenerative diseases has shifted to axonal survival (or sparing) rather than strictly neuronal survival and the prevention of apoptosis. Administration of Wld^S protein, or portions of it, could potentially act as a therapeutic agent to delay degeneration as shown in models of Parkinson's and motor neuron disease [16, 49, 53–55]. In the case of diseases, slowing degeneration may have significant benefits as opposed to injury where it could inhibit regeneration.

2.2.4   ROLE OF GLIA IN DEGENERATION AND REGENERATION OF CNS AXONS

Owing to different extracellular milieu, CNS neurons have a more difficult time regenerating than in the PNS. Schwann cells (SCs) are an important component of PNS regeneration, and act with macrophages to breakdown myelin and form new sheaths to guide the axon back to its target [56]. SC sheaths, termed bands of Büngner, are important for isolating the axon and growth cone from the damaged environment. Response to disease and injury in the CNS is quite different, since the guiding glial tubes that protect axons from surrounding environment are not present. Oligodendrocytes do not clear inhibitory myelin debris in the CNS, and astrocytes form a glial scar that permanently blocks passage of regenerating growth cones (Fig. 2.4A) [56, 57]. Microglia in the CNS are not nearly as efficient at clearing axon fragments as SCs and macrophages in the PNS, which can be detrimental to the nerve stump attempting to regenerate.

2.2.5   OLIGODENDROCYTES AND MYELIN ASSOCIATED INHIBITORS

Some membrane proteins, as well as fragmented and intact myelin from oligodendrocytes, are capable of inhibiting neurite outgrowth of CNS neurons. The Nogo family of membrane proteins is known to inhibit axon growth, and two specific inhibitory domains have been identified in Nogo-A that can both be detected on the extracellular surface of oligodendrocytes [58]. A recent study used gene silencing to knockdown Nogo receptors in transplanted neural stem cells (NSCs), which led to increased functional recovery in rats with TBI over cells transplanted without the receptor silencing [59]. The Nogo family of proteins exhibits complex interactions with axons and is discussed in further detail in Chapter 5 [14, 60]. Myelin-associated glycoprotein (MAG) has been shown to not only inhibit neurite outgrowth, but to initiate growth cone collapse [61]. Although MAG has so far proved detrimental to axon regeneration, it is known to encourage embryonic neurite outgrowth and has been implicated in maintaining and encouraging healthy, myelinated axons [62].

2.2.6   ASTROCYTE ACTIVATION AND GLIAL SCARRING IN THE CNS

Membrane proteins from oligodendrocytes and myelin are not the only source of inhibition at CNS injury sites. Microglia and astrocytes are recruited to damaged CNS areas, and while some astrocytes may support axons, injury stimulates a reactive phenotype characterized by cell hypertrophy. In response to injury, engorged astrocytes upregulate glial fibrillary acidic protein (GFAP) and vimentin expression in response to cytokines and growth factors released by microglia and other immune cells [63]. Vimentin and GFAP have both been shown to negatively affect axon regeneration [64]. Reactive astrocytes also upregulate CSPG expression, which inhibits regeneration at high concentrations. At the site of injury the concentration of CSPGs is very high but decreases as distance from the center of the insult increases [65]. The mechanism of inhibition from CSPGs is still not completely clear; some evidence suggests the GAG side-chains are to blame while other research points to the protein core. Treatment of injured sites with chondroitinase ABC (enzyme that cleaves GAGs from proteoglycans) decreases inhibition and could thus be used therapeutically to aid regeneration [57]. In the event of injury, the permeability of the blood-brain barrier and blood-spinal cord barrier is affected, resulting in infiltration of macrophages and cytokines that induce an inflammatory response [66–68]. Ultimately, astrocytes play a protective role in reestablishing these barriers and preventing the spread of injury; however, the effect is a highly inhibitory environment, caused by glial scar formation and release of neurodestructive molecules, resulting in failed neuron resynapsis and thus permanent loss of CNS function [69].

CHAPTER 3

Biomaterials for Scaffold Preparation

3.1 DEFINITION OF BIOMATERIAL AND REQUIREMENTS FOR NEURAL TE SCAFFOLDS

Most TE approaches (Fig. 1.1) begin with a biomaterial scaffold of natural or synthetic origin to provide structure for cells while preventing cavitation caused by massive tissue loss. Although the exact make-up of an ideal CNS TE construct is rarely agreed upon, some general requirements are widely accepted. A desirable scaffold is [70, 71]:

- biodegradable with the ability to release therapeutic agents if necessary

- mechanically similar to target host tissue

- easy to manufacture and process

- adhesive for cells or can easily be modified to be cell adhesive

- biocompatible or elicits an appropriate host response minimizing inflammatory and immune reactions

The ECM is extremely important for normal cell behavior and tissue function [72]. A common approach in TE scaffold design is to mimic the natural environment to facilitate full regeneration of damaged tissue. Often this includes using different materials to recreate the ECM. Surface interactions of biomaterials with both endogenous and exogenous tissue are extremely important due to the dependence of cell adhesion and migration to cellular functions. Figure 3.1 exhibits the different reactions of CNS host tissue to the same material with varied surface charge [73]. Chemical or physical surface modifications can be used as a strategy to remedy the unfavorable tissue interactions of some materials.

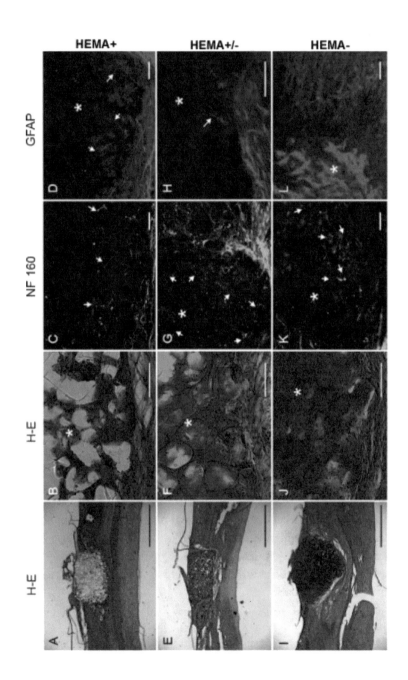

Figure 3.1: *Continues.*

Figure 3.1: *Continued.* Image of pHEMA gels implanted into spinal lesions, illustrating the importance of surface chemistry on tissue ingrowth. Gels with negatively, mixed, or positively charged surface were tested. Tissue penetration is shown in A, B, E, F, I, and J by hematoxylin-eosin stained sections. C, G, and K sections are stained for neurofilament and display the increased axon infiltration in the scaffold on the mixed charge surface. In D, H, and L sections, GFAP staining shows astrocytic activation of the positively charged pHEMA scaffold. Scale bar A, E, I = 1 mm; B, F, J = 50 μm; C, D, G, H, K, L = 100 μm. Image reprinted from [73].

An introduction to materials and common modifications will accompany specific material examples and subsequent chapters will carry these topics forward. Materials discussed here can be found in Table 3.1; in depth examination of these biomaterials and others can be found in review papers dedicated to biomaterials [70, 74–77].

3.1.1   BIODEGRADABLE SCAFFOLDS

Non-degradable biomaterials are used in neural TE; however, utilizing a material that can degrade over time and be completely replaced by natural tissue is typically the preferred approach. Most non-degradable materials offer control of synthesis and less complex design considerations, but at the cost of becoming a permanent fixture in the body [76]. Degradable scaffolds eradicate the necessity for surgical removal and allow for complete reinstallation of host function. In addition, when materials are designed to be degraded and eradicated from the body, they do not induce lasting immune or inflammatory responses. In the body, implanted materials can chemically degrade by enzymatic mechanisms or by hydrolysis. Cell interactions with the surrounding matrix are important for homeostasis, allowing for endogenous or exogenous cell migration throughout the area during regeneration [72]. The polymer on its own must be harmless to the body; in addition, its monomers should be nontoxic during the length of degradation time without eliciting any response preventing it from fulfilling its purpose. This often requires preventing significant alteration to the local cellular environment (*e.g.*, pH, osmolarity, microglial or astrocytic activation). For example, poly(lactic-co-glycolic acid) (PLGA) degrades into lactic acid and glycolic acid, which can decrease the local tissue pH, encouraging inflammatory responses [74]. Slowed degradation could circumvent pH changes, as long as degradation products are removed from the local environment quickly.

Most naturally derived biomaterials, such as proteins or polysaccharides, can be degraded by enzymes, and do not have toxic byproducts since specific elimination mechanisms exist in the body. In particular, proteolytic enzymes are integral to the processes of tissue remodeling and formation where migrating cells require active control of the ECM. Matrix metalloproteinases (MMPs) have been identified as important proteases in cell migration and ECM remodeling [78, 79]. The study of MMP proteolysis has lead to the identification of specific peptide sequences or substrates for each MMP [80–82]. These peptide substrates are short (< 7 amino acids) and have been incorporated into polymeric crosslinkers in biomaterial scaffolds [78, 79, 83, 84]. The MMP-1, or collagenase, cleavable sequence Ala-Pro-Gly-↓-Leu (↓ for cleavage site) has been incorporated into

Table 3.1: Common biomaterials used in CNS tissue engineering.

Material	Chemical Formula	Source	Degradable	Form	References
Agarose		Red algae	No	Preformed gel, injectable gel, electrospun mesh	[97, 115-116, 498]
Alginate		Brown algae	Dissolution at neutral pH	Preformed gel, electrospun copolymer	[75-76, 118-120, 129]
Chitosan		Crustaceans	Enzymatic	Preformed gel, injectable gel, electrospun copolymer	[94, 121, 123, 129-130]
Collagen		Animals, humans	Enzymatic	Preformed gel, injectable gel, electrospun mesh	[74, 76, 106-107, 111-112]
HA		Animals, humans	Enzymatic	Preformed gel, electrospun mesh	[75-76, 91-92, 99, 109, 114]
Matrigel		Mouse sarcoma	Enzymatic	Preformed gel, injectable gel, electrospun mesh	[75-76, 133]
Methyl-cellulose	R=H or CH3	Plants	Enzymatic	Preformed gel, injectable gel	[71, 75, 91, 99]
PCL		Synthetic	Hydrolytic	Electrospun mesh	[105]
PEG		Synthetic	No	Preformed gel	[75-76, 134-135]
pHEMA		Synthetic	No	Preformed gel	[73, 136]
PLGA		Synthetic	Hydrolytic	Preformed gel Electrospun mesh, drug delivery	[91, 105, 135]
PPy		Synthetic	No	Film, particles Preformed ge	[40-45]
SAP		Synthetic	Enzymatic or hydrolytic	Injectable gel	[76, 137-138]

the backbone of photopolymerizable poly(ethylene glycol) (PEG) allowing MMP-1 mediated cell migration through the hydrogel scaffold [78, 79]. The MMP-2, or gelatinase, peptide substrate Pro-Val-Gly-↓-Leu-Ile-Gly (PVGLIG) has been used as a linker for dextran and methotrexate to enable the creation of MMP-2 activated drug delivery microparticles [84]. The PVGLIG peptide has also been included in larger self assembling peptides for the creation of MMP-2 sensitive self-assembled hydrogels. The gelatinases (MMP-2 and 9) are important in neural development and differentiation [85]. MMP-2 has been shown to be particularly important in postnatal development and in migration [86] and axon outgrowth [87]. In the native CNS environment, MMPs facilitate migration and differentiation; thus, CNS tailored biomaterial scaffolds should allow for material remodeling to promote differentiation, migration and cell process extension.

In contrast to enzymatic degradation, hydrolytic degradation occurs at hydrolytically labile bonds, such as esters, orthoesters, and anhydrides. Due to the patient-to-patient variation in autogenous enzymes, hydrolytic degradation of materials is more predictable over time and location, while holding standard throughout different populations of patients, as compared to enzymatic degradation [88, 89]. Polymers degrade via bulk or surface erosion mechanisms. Bulk deterioration of hydrolytically degraded polymers occurs over the whole volume due to faster water penetration as compared to hydrolytic cleavage of bonds at the surface [70]. Surface erosion of a polymer occurs when water does not penetrate easily and bonds are cleaved before water can reach the inner mass of the material. This type of degradation is characterized by mass loss at the surface that penetrates inward over time. Most often, surface erosion leads to a more linear release of any encapsulated agents included in the scaffold [70, 90]. The mechanism and rate of degradation are typically selected depending on the application. Enzymatically degradable materials are incorporated more often into TE constructs and hydrolytically cleavable materials are typically used to achieve stable release profiles in drug delivery applications. An example of a scaffold with both types of degradation is examined later, where an enzyme susceptible scaffold is impregnated with a nanoparticle delivery system with the potential to release therapeutic drugs or growth factors [91].

3.2   SCAFFOLD CREATION STRATEGIES

3.2.1   HYDROGELS

Hydrogels are a very popular choice for neural TE scaffolds due to their characteristic similarity to CNS tissue. Hydrogels are polymer networks typically composed of 1-5 wt% polymers that swell with water and can have a low mechanical stiffness, similar to native soft tissues [70, 71]. Mechanical, as well as surface, properties have been shown to be important for neural cell adhesion, migration and survival, as well as the differentiation of stem cells [92–95]. To control the degree of crosslinking, and indirectly the mechanical properties, the polymer can be chemically crosslinked, photopolymerized, or irradiated either in a dry state or in solution [70, 75]. Photoinitiation of crosslinking in gels is the preferred approach to fill lesioned cavities of irregular shape and size, especially in neural TE where reduced chemical linking is required for soft gel formation. Alternatively, preformed gels can use any crosslinking method and be shaped accordingly. Swelling behavior of a hydrogel influences properties such as nutrient/waste diffusion, surface properties, optical properties, and mechanical properties [70]. Diffusion of nutrients and waste products throughout the scaffold is essential for cell maintenance within the scaffold and for encouraging the ingrowth of homologous cells and tissues when the construct has been incorporated into the body. One way to control swelling behavior is through environmentally sensitive hydrogels. The most common environmental cues are pH and temperature [70]. Researchers have utilized environmentally sensitive hydrogels that exhibit complexing in the presence of a stimulant to initiate physical crosslinking from inter or intrapolymer interactions. For this reason, thermogelling polymers have become popular because they can be injected, and crosslink into place at body temperature, obviating the need for invasive surgical implantation [91, 96–99]. Overall, hydrogels are the most common form of CNS scaffolds, and have been widely employed to fill nerve guidance tubes for PNS constructs [100, 101].

3.2.2   ELECTROSPUN FIBERS

Electrospinning has been adapted for a variety of regenerative medicine applications, including neural TE due to its ease of use, versatility in fiber thickness and composition, and characteristic similarity to ECM on the nanoscale [102, 103]. The relatively simple set-up for electrospinning is shown in Figure 3.2, which consists of a voltage generator, a spinneret, and a grounded collecting plate. The process requires that a voltage is applied to a liquid droplet at the tip of the spinneret causing a jet of fluid to travel to a collecting plate [102, 104]. As the voltage is applied to the droplet, it works against surface tension initiating the formation of a Taylor cone, where a stream of liquid is ejected from the droplet towards the collecting plate. A pump is typically used to keep the polymer solution at a constant flow rate to the end of the spinneret, and is the simplest parameter that can be adjusted to control fiber size. In addition to feeding rate, the collection gap, electric field strength, polymer, and solvent can be varied to adjust the fiber diameter. Electrospinning on a normal collecting plate yields a mesh of randomly aligned fibers; however, further modifications of the collecting apparatus can be made to create parallel fibers. Rotating mandrels and drums, oscillating frames, metal frames, and a pair of electrodes with an insulating gap have been used to acquire specifically aligned electrospun fibers (Fig. 3.2) [104]. Researchers have also demonstrated ways to tune electrospun fiber morphology and composition by altering the collecting method. One distinct method used is a collecting media, as opposed to the traditional collecting plate, which allows for porous fibers or for the incorporation of different therapeutic agents into distinct fiber regions [104]. Electrospinning offers a relatively easy and cheap technique to create highly tunable TE scaffolds. Specific examples of electrospinning for neural TE will be discussed later; however, it should be noted that hydrogels are still more common than electrospinning for the formation of CNS constructs and newer hybrid approaches of the two offer advantages of both techniques [105].

3.3   CURRENT BIOMATERIALS IN CNS TE

There are many different biomaterials that have been used in CNS regeneration, and those discussed here are not an exhaustive list. The following discussion should serve as an introduction that is

Figure 3.2: (Top) Electrospinning set-up displaying the syringe pump, syringe needle, voltage generator, and stationary collecting plate. Inset shows fiber of the electrospun polymer traveling through the electric field before it reaches the collector. (Bottom) Electrospun meshes can take on a variety of forms (all fibers made from PCL). (A) Random fiber meshes are easy to form with a stationary collecting plate. (B-C) Specifically aligned fibers can be made using rotating mandrels or spinning collectors, and layered scaffolds can be made by repeating the process. (D) Another method used to form aligned fibers involves two electrodes and an insulating gap between. Insets show random fibers on the electrodes (left inset) and parallel fibers formed between the electrodes (right inset). Image reprinted from [104].

designed to provide the reader with a basic understanding of common neural TE biomaterials, with specific examples of past applications and unique features of each material. Reiterated throughout the literature is the commonality that a successful TE construct requires a composite of many stimuli built into the scaffold. Many of the materials discussed here are typically used as copolymers, blends, or in conjunction with other molecules to encourage the appropriate tissue responses both *in vitro* and *in vivo*.

3.3.1   NATURAL MATERIALS

Many researchers have used the body's most abundant ECM component, collagen, to form constructs for neural TE. Collagen is composed of chains of amino acids linked by peptide bonds, or the covalent bond between the amino group of one amino acid and the carboxylic acid of another amino acid. These chains form secondary structures and tertiary structures; combinations of multiple tertiary structures form higher order structures and finally collagen fibrils [74]. Over 20 collagens have been identified; however, collagens I-IV are the most abundant in the body and only not all collagen types are found in the brain [5, 74]. Since collagen is native to the body, it is recognized by cells, promoting adhesion, however, with the potential for immune response. For this reason, allograft or xenograft collagens are especially avoided due to their immunogenicity and possibility of disease transfer [74, 106]. Enzymatic degradation of collagen is carried out by MMPs at specific cleavage sites. As previously noted, collagen I can be cleaved by MMP-1 to allow for matrix remodeling and cell migration. MMP-2 and -9, both gelatinases, can degrade collagen IV, a common basement membrane protein found in the CNS [85]. As an insoluble protein, collagen is most often dissolved in a cold, acidic solution and can be made into a hydrogel by adjusting the pH and temperature to physiological conditions [76]. In this way, collagen can be used as an injectable hydrogel for TE applications. In contrast to physically formed gels, chemical crosslinking of collagen (by aldehydes, carbodiimides, polyepoxides, hexamethylene-diisocyanate, etc.) is often used to control scaffold properties such as the mechanical properties, degradation time, and immune reactivity [74, 106]. Concentration and the degree of crosslinking can be used to adjust the stiffness of collagen gels, as discussed earlier; collagen can also be blended with other materials to increase mechanical compliance or stiffness, depending on the additive. As a native ECM component, collagen has integrin binding sites, and can be further enhanced for neuronal regeneration by incorporating growth factors such as neurotrophins. Neurotrophin-3 (NT-3) was recombinantly functionalized with a collagen binding domain (CBD), allowing it to complex to collagen I scaffolds; these scaffolds were subsequently implanted into the transected spinal cord of rats with positive functional results [107]. The collagen-NT-3 combination yielded significant locomotor improvement over collagen scaffolds with soluble NT-3. NGF was also functionalized with a CBD and used as an injectable scaffold for sciatic nerve regeneration (discussed later in Chapter 5) [108]. Gelatin, a hydrolyzed derivative of collagen, can also be used in neural TE to make gel scaffolds [109, 110]. Collagen can be processed into gels, meshes, and even a powdered form to suit a variety of needs. In addition, collagen can be electrospun into a fibrous mesh and crosslinked for stability, leading to excellent neural outgrowth responses *in vitro* [111, 112].

Since it is a regular component of the ECM, HA (hyaluronic acid) is being studied as a biomaterial for the CNS [70, 75]. HA is very viscous due to its high molecular weight and can form soft gels via chemical crosslinkers, such as carbodiimide, or modified for photocrosslinking [76, 113]. Natural degradation of HA occurs in the body because of interaction with free radicals, MMPs, and hyaluronidases allowing for native remodeling of these scaffolds [74]. Increased spreading and attachment of hippocampal neurons on HA gels as compared to unmodified HA gels were achieved with immobilized Nogo receptor antibodies (antiNgR) [92, 114]. Further, HA-antiNgR gels were able to support NSC survival and differentiation [92]. HA on its own has poor mechanical properties; thus, it is often blended with polymers and stiffening agents to increase rigidity. HA has been used with collagen and methylcellulose to decrease their compressive elastic moduli closer to $\sim$1 kPa, which is similar to brain tissue [99, 113]. HA may be preformed or injected to form brain and spinal TE constructs.

Agarose is a thermo-setting, linear polysaccharide isolated from red algae, and is often incorporated in the food industry for its gelatinous properties. Agarose can be formed into a hydrogel by dissolution with heat and subsequent cooling until gelation; however, traditional agarose gels must reach a temperature below physiological body temperature to fully set and thus require a cooling system to gel *in situ* [97]. Different types and derivatives of agarose are now commercially available that can gel at many temperatures, including close to body temperature, to alleviate the need for complex cooling systems. The mechanical properties and porosity of agarose can be varied by adjusting its concentration in the initial solvent. Agarose has been shown to exhibit minimal inflammatory response; however, it is undesirably non-biodegradable and does not allow cellular adhesion [115]. Agarose hydrogels can be molded into a variety of shapes in order to treat many different sized and shaped lesions. A channeled agarose scaffold containing bone marrow stromal cells (bMSCs) and NT-3 induced spinal cord axons to penetrate into the channels with good linear alignment due to the construct's honeycomb geometry (Fig. 3.3) [116].

Alginate is an anionic linear polysaccharide extracted from the cell walls of brown algae, which gels in the presence of multivalent cations such a Ca^{2+} [76, 117]. Alginate can be formed into hydrogels, and has recently been used in cryogels [118]. To form a cryogel, the monomers are gelled at subzero temperatures, allowing ice crystals to form large, interconnected pores when the gel is later thawed. Alginate/agarose cryogels show excellent mechanical properties and cellular attachment; however, they were not tested with neuronal or glial cells [118]. Although it cannot be electrospun alone, alginate can be blended with other polymers to create fibrous structures. Alginate/poly(ethylene oxide) (PEO) meshes can support *in vitro* fibroblast attachment, which can be further enhanced by the immobilization of adhesive peptides [119, 120].

Chitosan is derived from chitin, a naturally abundant polysaccharide found in the shells of crabs and other shellfish. Considered biocompatible by most researchers, chitosan may illicit an inflammatory response through macrophage activation with low deacetylation; therefore, it is typically

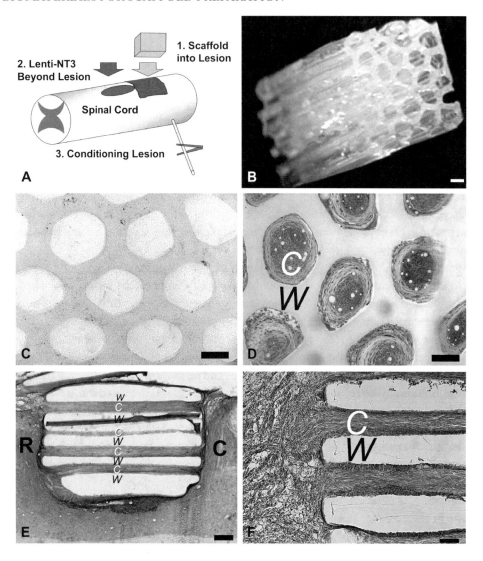

Figure 3.3: Templated agarose scaffolds. (A) Schematic of implantation of agarose scaffold (1) that could be accompanied by lentiviral NT-3 delivery (2), and/or conditioning lesions of the sciatic nerve (3). (B) Macroimage of agarose scaffold showing channels in a columnar configuration. (C) Cross-section of agarose scaffold where honeycomb like pores are seen. (D) Cell infiltration seen in the channels (white C) after four weeks of implantation. Scaffold walls (black W) are still intact. (E, F) Sagittal view of scaffold where tissue integrated into the length of the scaffold; however, there is a slight reactive cell barrier surrounding the implant. (black R, rostral; black C, caudal; white C, channel; black W, wall or scaffold). Scale bars B = 200 μm, C = 100 μm, D = 250 μm, E, F = 100 μm. Image reprinted from [116].

used at deacetylation percentages above 80% [121]. Chitosan is recognized to have antibacterial properties, which are a useful property for biomaterial scaffolds [122]. Chitosan can be covalently bonded or thermogelled in the presence of glycerophosphate salt or blending with a thermogelling polymer. Thermogelled chitosan was able to support comparable mouse cortical neuron survival and neurite outgrowth *in vitro* to cells grown on poly(D-lysine), a common adhesion enhancing agent [123]. Alternatively, a methacrylated form of chitosan has been synthesized that can be formed into hydrogels via exposure to UV light in the presence of a photoinitiator [124]. This methacrylated chitosan can support the survival and differentiation of cultured NSCs and allows for the easy tuning of substrate stiffness via photoinitiator concentration or exposure time [94, 124]. Chitosan can be formed into a cryogel, either alone or blended with other polysaccharides, similar to alginate and agarose [125–128]. In addition, chitosan blends have the capacity to be electrospun, and these scaffolds have been used in PNS TE [129, 130].

Methylcellulose (MC) is an inverse thermogelling polymer that is derived from cellulose (a polysaccharide found in plant cell walls), and has been shown to invoke a low inflammatory response *in vivo* [71, 75]. Due to its hydrophilicity, MC has low cellular adhesion unless modified, but shows biocompatibility *in vitro* and *in vivo* [131, 132]. An injectable hydrogel composite of HA and methylcellulose (HAMC) incorporating synthetic polymer nanoparticles showed very little microglial response when injected into the intrathecal space of rat spinal columns [91]. The HAMC combination allowed for fast gelling at physiological temperature [99]. Blank nanoparticles were included as a test for a drug delivery vehicle that could release therapeutic agents in future studies [91].

Matrigel™ (BD Bioscience) is a thermogelling mixture of ECM proteins isolated from Engelbreth-Hom-Swarm mouse sarcoma. The complex medley of proteins leads to intricate cell interactions and behaviors which have been shown to often be favorable for neural regeneration; however, due to its derivation from mouse tumor cells, it is unlikely that Matrigel™ can be used for human implantation [75, 76]. Similar to use of fetal serums, Matrigel™ is not necessarily applicable to translation to human studies, but successes using both suggest the need for multi-component systems with multiple cues. In addition to creating traditional scaffolds, Matrigel™ has been electrospun in an attempt to create constructs resembling basal lamina for enhanced SC growth [133].

3.3.2   SYNTHETIC MATERIALS

PEG is a hydrophilic polymer that is biocompatible and typically nonfouling and therefore often requires modifications to increase cellular adhesion [75, 76]. One way to adjust the surface interactions of PEG or its degradation characteristics is through polymer blends or composites. Bjugstad *et al.* recently performed a comprehensive assessment of the biocompatibility of PEG hydrogels in the brain. In the study, PEG/lactic acid (PEG/LA) gels were injected into rat cortexes, with the slow degrading (less LA) and non-degradable (no LA) PEG gels demonstrating the lowest microglial and astrocytic response [134]. The non-degradable and slower degrading gels showed less glial activation in a 50-200 μm region surrounding the implant than the sham. While LA is useful for making a PEG based scaffold degradable, it likely leads to observed glial activation by increasing the acidity in the local environment. In other studies, photopolymerized LA/PEG hydrogels were utilized as a delivery system for the neurotrophic agent NT-3 [135]. When the scaffolds were implanted into hemisectioned rat spinal cords, improved movement and coordination, axon ingrowth, and host NT-3 concentrations were observed.

Poly(caprolactone) (PCL) is a polyester with high solubility, low toxicity, biodegradability, and the capacity to blend with other materials [70]. Ester bonds on the PCL backbone allow it to be hydrolytically degraded. Recently, PCL/PLGA electrospun hollow tubes were filled with self-assembling peptides and implanted into spinal contusion cyst sites, leading to endogenous cell ingrowth but little functional recovery [105]. Results from PCL/PLGA electrospun fiber testing in the CNS have been positive as well, showing the diversity of both polymers.

Methacrylate gels are widely used in biomedical engineering. In particular, poly(hydroxyethyl methacrylate) or pHEMA, which is well known for its use in soft contact lenses, has shown utility for CNS TE. Although pHEMA is non-biodegradable, it has many desirable qualities including its inertness and ability to be heat sterilized; pHEMA is easy to prepare and allows for the tuning of its mechanical properties with the adjustment of crosslinking density [70]. As mentioned in the introduction to Chapter 3, surface chemistry can be modified to adjust cellular interactions with pHEMA gels (Fig. 3.1). Mixed charge and positively charged gels demonstrated the best connective tissue penetration, but the mixed charge functionalized gels had a lower astrocytic response [73]. Yu *et al.* produced a co-polymer scaffold of pHEMA with 2-aminoethyl methacrylate (AEMA) to allow the attachment of laminin-derived peptide sequences to increase neuron adhesion to HEMA scaffolds [136]. Primary sensory neurons adhered and sent out longer processes on the peptide modified gels over unmodified gels; and cell areal cell density was eight times higher as assessed by fixation and immunostaining [136]. Both surface modifications of the methacrylate gels showed improved cellular response, and are common to many scaffold strategies used in CNS TE.

Self-assembling peptides (SAPs) offer an attractive injectable biomaterial scaffold for spinal cord and brain injuries since they form under physiological pH and temperature. Engineered peptides allow a defined linear chain of repeating units of amino acids to be produced synthetically and are easily processed [76, 137]. Scaffold mechanical properties and assembly are dependent on the characteristics of the peptides used. High hydrophobicity and longer amino acid sequences promote scaffold assembly with better mechanical properties [138]. One advantage of utilizing SAPs is that they can be functionalized to promote cellular activities such as migration and differentiation. In a recent study, the self-assembling 16 amino acid peptide RADA16-I was enhanced with a bone marrow homing motif to increase cell survival within the injured spinal cord of rats, and it was shown to significantly improve motor control and coordination after 7 wks of recovery (as evaluated by the BBB scale) [137]. In this study, RADA16-I was delivered by 3 injections into the lesion and gelled via salt induction [137, 138].

Since electrical activity is an important property of neural tissue, conductive polymers are also used in neural TE constructs. Conductive polymers have loosely held electrons in their backbones and use charge transfer from dopant molecules to enter a highly conductive state [139]. Poly(pyrrole)

(PPy) and poly(aniline) (PANI) are the most common conducting polymers and have been used in neural applications. Conducting films of ester functionalized PPy with conjugated NGF supported the culture of PC12 cells *in vitro*, displaying the biocompatibility of PPy as well as its ability to covalently attach growth factors while remaining conductive [140]. Biodegradable electrically conductive composites have been created by mixing PPy nanoparticles into hydrolytically cleavable scaffolds. PPy/polylactide composites are the most popular choice for scaffolds and have been tested with fibroblast and neuronal cell cultures [141–144].

Neural TE scaffolds exhibit strict demands on their bulk biomaterial properties as well as surface characteristics. The materials discussed here all possess strengths and flaws, which should be weighed against another depending on the specific application. In many studies, natural and synthetic materials can be blended or conjugated together for improved properties and to gain additional functionality. For CNS TE, it is important to remember the restrictive environment that can be created by microglial and astrocytic activation, and to make sure any material used in a TE scaffold does not illicit any undesirable response. Subsequent discussions in Chapter 5 will also provide insight into further surface modifications, physical and chemical, that can mask bulk material properties and further enhance cellular responses of the scaffold material.

CHAPTER 4

Cell Sources for CNS TE

When designing CNS tissue constructs, inclusion of encapsulated and surrounding cells are of great importance to achieve tissue regeneration (Fig. 1.1). Biomaterial scaffolds discussed previously are commonly impregnated with cells to expedite recovery by replacing lost tissue or by decreasing the migrational distance of natural cells in the surrounding area. In this chapter, a brief overview of commonly used and promising cell sources follows, and is meant to give the reader an introduction to each one. Table 4.1 presents common cell sources and their acronyms, and offers references for further reading.

Table 4.1: Common cell sources for CNS tissue engineering.

Cell Type	Abbreviation	Potency	Source	Reviews
Embryonic stem cells	ES cells	Pluri	Embryo	[175, 499-500]
Induced pluripotent stem cells	iPS cells	Pluri	Multiple: skin, embryo, SVZ	[201, 501-502]
Mesenchymal stem cells	MSC	Multi	Bone marrow, umbilical cord, adipose tissue	[251-252]
Skin-derived precursors	SKP	Multi	Skin	[263, 503]
Neural stem cells	NSC	Multi	Multiple: SVZ, DG, spinal cord	[221, 504-505]
NSC: Dentate gyrus cells	DG cells	Multi	Hippocampus, specifically the dentate gyrus	[221]
NSC: Spinal cord ependymal cells	Ependymal cells	Multi	Spinal cord	[23, 221, 506]
NSC: Subventricular zone stem cells	SVZ cells	Multi	Cortex, walls of the lateral ventricles	[221, 507-508]
Olfactory ensheathing cells	OEC	None	Nasal cavity	[509-510]
Schwann cells	SC	None	Peripheral nerves	[508, 511]

TE design criteria for cells are similar to those for biomaterials, in that any elicited response that impedes regeneration is unacceptable. For this reason, autologous cells are the gold standard for TE since they rarely evoke an undesired response (there are exceptions, discussed later). However, the price of using the patient's own cells is paid by donor site morbidity, as well as additional time spent on surgical isolation and culture. In addition, care must be taken to correctly choose cells for different injury and disease states. For demyelinating wounds and diseases (e.g., MS), reestablishment of the myelin sheath by glial cells is of utmost importance. In contrast, when neurons or both neurons and glia are lost, typically strategies shift focus to the neuron. Subsequent discussion of specific cell types will show that opinions differ on the optimal therapy in these cases. Some researchers attempt neuronal implantation while others incorporate glia or stem cells to encourage native neurons to regrow through the lesioned area. An overall strategy for cell incorporation or stimulation is not agreed upon, and is usually formulated for the specific application.

Somatic cells of the CNS are rarely used as a cell source for CNS TE. Issues of secondary injury sites, surgical accessibility, and poor mitotic ability restrict the use of primary cells. As an alternative, the use of the peripheral myelinating cells have become popular; this includes protective ensheathing cells within the nasal cavity. During homeostasis and development, glia serve as the supporting cells of the neuron. Most CNS TE strategies incorporating glia focus on maintaining unaffected surrounding tissue while remyelinating damaged tissue.

Stem cells offer an attractive alternative to somatic cells, and provide increased cell expansion and decreased donor site morbidity. Stem cells can be derived from a number of sources, and depending on their location and age, they possess different variations of two intrinsic stem cell characteristics: differentiation (or ability to form multiple cell types) and proliferation (or ability to self-renew). Since they can be readily expanded, less stem cells are needed initially to achieve high cell numbers compared to somatic cells. Totipotent stem cells have the ability to become any cell type of the body, as well as expand indefinitely; from here differentiation potential or plasticity decreases to pluripotent, to multipotent and finally to progenitor stem cells. Pluripotency is possessed by cells very early on in development, namely embryonic cells, but also has been induced in somatic cells by genetic alteration, as will be discussed later this chapter. Multipotent and progenitor stem cells persist throughout the body in stem cell niches into adulthood, and several populations of adult NSCs reside within the CNS.

4.1   PRIMARY CELL TREATMENT OF CNS INJURY

4.1.1   GLIAL CELLS

As mentioned previously, SCs (Schwann Cells) are the myelinating cells of the PNS. SCs wrap around the axon many times allowing multiple SCs to myelinate a single axon. In contrast to oligodendrocytes in the CNS, SCs resorb fragmented myelin and the distal axon from an injured site, recruit macrophages, and then work in synergy to clear debris and encourage axon resynapsis [145, 146]. SCs play an encouraging role in PNS regeneration and serve to augment axon synapse reformation. They do this by transforming to a non-myelinating phenotype, increasing secretion of axon recruitment factors brain derived neurotrophic factor (BDNF, sometimes referred to as BDGF), ciliary neurotrophic factor (CNTF), and NGF, and forming a tube that guides the growth cone of the proximal axon to its destination [147, 148]. Though not typically found in the CNS, SCs are capable of myelinating CNS neurons; and endogenous SCs have been found to migrate into the injured spinal cord and myelinate CNS axons [149–151]. The mechanism of SC migration into the CNS is still unknown, as is the extent of their participation in CNS nerve regeneration, though it is known that their contribution is minimal [152, 153]. Encouragement of SC migration into the injured spinal cord has not been well researched; rather, the majority of studies to date have injected or transplanted SCs into the injured area in an attempt to encourage regeneration. SC myelination of axons is limited to regions with low astrocyte numbers. *In vivo* and *in vitro*, astrocytes and SCs are known to inhabit mutually exclusive areas. Early studies by Blakemore *et al.* created areas of demyelination in the cortex or spinal cord with ethidium bromide, which resulted in low or compromised astrocyte populations [154, 155]. SCs were able to myelinate endogenous axons in the areas with decreased astrocytic inhabitance. Modifications to SCs using exogenous proteins or transcription factors increase SC interactions with astrocytes and improve their migration and integration into the CNS. Recent work has focused on altering SC neural cell adhesion molecules (NCAMs) by inducing SC expression of polysialic acid (PSA) through viral delivery of sialyltransferase X (STX) [156, 157]. PSA associates with the fourth domain (extracellular portion) of NCAM and decreases adhesiveness so that the cells can more easily separate. Expression of PSA-NCAM is found on oligodendrocyte precursors during development and regeneration, and has been found to increase SC migration into astrocyte territory without adversely affecting their myelination capabilities [156]. Adult primate SCs were transplanted near experimentally demyelinated areas of the spinal column and demonstrated faster and more efficient migration when transfected with STX viral vectors [157]. In this study, remyelination was enhanced in transfected SCs; however, no functional analyses were conducted. One significant advantage of SCs is their ease of isolation and expansion capabilities. Obtaining SCs by biopsy, expanding them in culture with mitogenic agents, and purifying them from fibroblasts provides a source for autologous cells [158–160]. Glial growth factor (GGF) and the cAMP activator forskolin are two popular mitogenic agents, but as with any mitotic factor, their use must be closely monitored and inhibited before implantation to avoid tumorigenesis [158, 159, 161, 162].

Olfactory ensheathing cells (OECs) are radial glia of the olfactory bulb and nasal mucosa, located in the upper region of the nasal cavity. OECs are similar to SCs in their CNS regenerative ability, with the exception that they are less inhibited by the presence of astrocytes [163, 164]. OECs have been used in sensory and motor tract regeneration studies, but results are mixed and it is unclear whether they enhance tract regeneration or simply increase axon sparing [164, 165]. The corticospinal tract is the group of motor fibers in the brainstem connecting the cortex with the spinal cord, which are responsible for transmitting signals required for skilled movements. In the corticospinal tract, transplanted OECs have facilitated positive recovery of forepaw motor skills in animal models [166, 167]. The motor tract responsible for large movements, i.e., the rubrospinal tract, has shown limited regenerative success using OECs, even with the addition of growth factors such as BDNF or glial cell-derived neurotrophic factor (GDNF) [168, 169]. Age of OECs is an important factor to consider for remyelination and functional recovery. OECs isolated from juvenile rats were able to migrate to post-transplantation lesions created with ethidium-bromide, which is not observed with the use of adult OECs [170–172]. Age is not the only factor affecting the efficacy of OECs. Source, location, and delay time between injury and therapeutic administration often play a significant role in effectiveness [165, 170–172].

In an attempt to utilize positive attributes of both cell types, OECs have been implanted in combination with SCs. OECs have been shown to enhance SC myelination capabilities as well as decrease astrocytic segregation of SCs [173]. Axon growth into SC bridges containing chondroitinase ABC enhanced by OECs at the interface between construct and endogenous cells [174]. This three-part combination was used in adult rat spinal cord lesions by Fouad *et al.* in 2005 and resulted in increased myelinated axons within the construct as well as serotonergic fibers re-entering the caudal spinal cord from the graft [174]. Increased myelinated axons correlated to increased functional recovery, as assessed by BBB scoring.

4.2   PLURIPOTENT STEM CELLS

4.2.1   EMBRYONIC STEM CELLS

Pluripotent stem cells, namely embryonic stem (ES) cells, have the ability to become neurons or glia for use in CNS TE when provided the correct cues. ES cells have also been proposed as an abundant source for NSCs, not only for use in cell based therapies but for screening assays and disease progression studies. It has even been argued that NSCs derived from ES cells retain better differentiation capabilities or 'developmental competence' [175].

One attractive property of ES cells for neural TE, other than their excellent proliferative capabilities, is their predictable behavior in response to developmental cues. Embryonic cells are derived from the inner cell mass of the blastula, and neuronal cell induction begins once the blastula enters gastrulation. Development of the neural system begins with induction of neural precursors which are differentiated to neurons and glia, proceeded by axon journey and preliminary synapse formation to their targets. Neural functionality is made possible by final remodeling of the neural network. Neural cells arise from the ectoderm, or the outermost layer of the gastrula, on the dorsal side where the inhibitory bone morphogenetic proteins (BMPs) are suppressed by the organizer molecules noggin, chordin, follistatin, cerberus and Xenopus nodal-related 3 (XNr3) protein (Fig. 4.1) [4, 176]. After the inhibition of BMPs, ectodermal cells are permitted to become neural precursor cells and are directed further by neural promoting molecules. Once the neural tube is formed in the embryo, spatial patterning occurs to direct cell fate. Caudal formation is led by Wnt-8, basic fibroblast growth factor (bFGF), and retinoic acid (RA) [177, 178]. The spinal cord is patterned ventrally by sonic hedgehog (Shh) protein and RA, and dorsally mainly by BMPs [179, 180]. The PNS is derived from neural crest cells that migrate from the neural tube; neural crest cells may become PNS neurons and glia, smooth muscle cells, or pigment cells [181–183]. The cell type of neural crest cells is determined

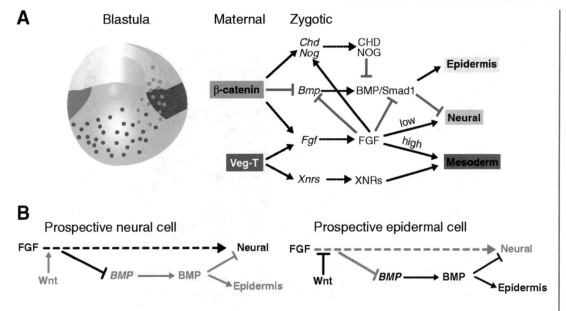

Figure 4.1: Diagram of neuronal development from the blastula in Xenopus (A) and chick (B), illustrating the roles of BMP, FGF and Wnt. For both species, FGF promotes neuronal induction by blocking BMP activity. Chd, chordin; Nog, noggin; NXR, nodal-related factors. Image reprinted from [494].

by migration route and environmental cues encountered. Events directing development have been well studied and are often mimicked by applying certain factors in a sequential manner to ES cells in order to derive NSCs and even specific somatic cell types [184–186].

Defined differentiation protocols eliminating the use of serum are in development to improve the clinical relevance of ES cells. In many cases, neural differentiation is initiated by culturing ES cells on feeder cells (typically embryonic mouse fibroblasts), followed by culture as suspended embryoid bodies with RA (Fig. 4.2). This culture regime has been found to promote a neuronal phenotype and suppress mesodermal cell types [187, 188]. Embryoid bodies are subsequently dissociated and plated on laminin coated surfaces with N2, a serum free neuronal supplement, in media for the neuronal phenotype [188]. Motor neuron specification has been achieved by activating the Shh pathway subsequent to RA; RA is necessary for early neural induction but should be followed by other factors if further differentiation is desired [184–186]. This differentiation regime mimics development closely, as ES cells are given caudalization signaling from RA and then sequential ventralization signaling from Shh. Testing of *in vitro* differentiated ES cells by most groups involves immunostaining or electrophysiology testing; however, Wichterle *et al.* alternatively implanted differentiated motor neurons into developing chick embryos to test functionality [186]. This group observed that ES derived cells localized into the correct area of the spinal cord and projected axons to the appropriate peripheral regions. Directing ES cell differentiation has not been limited to development-linked

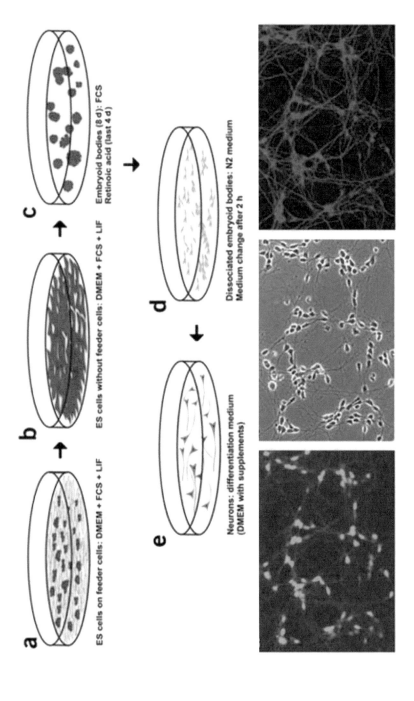

Figure 4.2: (Top) Schematic representation of neuronal differentiation of ES cells accompanied by actual images of the differentiated cells (Bottom). Left image shows endogenous GFP expression, center image shows the same field in phase contrast, and the right image shows βIII rubulin expression for the region. Image reprinted from [188].

factors and molecules. A recent study found that soluble amyloid precursor protein (APP) not only induced neural progenitor differentiation but also led to ES cell expression of βIII tubulin, a common neuronal marker, in only 5 d [189]. APP, a protein thought to be a synapse regulator, is more well known for its transformation into amyloid plaques in Alzheimer's disease [51, 52].

Since success has been achieved in directing ES cell fate into neural cell types, focus has shifted on incorporating ES cells or ES-derived cells into functional neural TE constructs. In an attempt to enhance serotonergic fibers two weeks after a hemisectioned SCI, embryonic neural precursor cells (NPCs) from the neural tube were injected caudal to the injury with generally positive results [190]. Even though the NPCs showed superior cell morphology and axon extension in tissue sections, functional improvement of coordination and maneuverability were comparable to injected mesenchymal stem cells (MSCs) and both showed significant improvement over control mice. Similarly, successful treatment of a simulated crush injury demonstrated the regenerative efficacy of human NPCs transfected to express Neurogenin 2 (Ngn2), a proneural factor that inhibits astrocytic differentiation, injected a week post-injury in adult rats [191]. NPCs grafted without Ngn2 did not result in any functional improvement; however, NPCs that expressed Ngn2 resulted in significantly higher scores in functional motor skill tests. This study by Perrin *et al.* showed the utility of human embryonic NPCs in a more realistic model where induced cells were administered a week after injury. Outside of the research lab, injury to CNS tissue is not treated immediately following insult due to delays in diagnosis, treatment availability, not to mention many other obstacles. The ability of human cells to perform well in animal models coupled with a more realistic delay of treatment makes this work very exciting for future CNS regeneration treatments.

4.2.2   INDUCED STEM CELLS

The ethical considerations surrounding ES cells and the associated controversy have dampened the enthusiasm for incorporating them as a cell source into TE strategies. In recent years, an alternative source of pluripotent stem cells is under active development. In 2006, Shinya Yamanaka's group demonstrated the creation of pluripotent stem cells from adult fibroblasts using gene therapy to introduce Oct3/4, Sox2, c-Myc, and Klf4 expression [192]. Yamanaka's group termed these cells, 'induced pluripotent stem (iPS) cells', and observed multiple lineage formation after injection into adult nude mice or blastocysts. Development of two factor iPS production (only Oct4 and Klf4) from mouse NSCs may offer better clinical relevance since it eliminates the c-Myc gene that is a well know oncogene implicated in tumorigenicity [193]. iPS cells behave similarly to ES cells and can be induced to form neurons in a similar manner (Fig. 4.3).

Despite the attractiveness of using the patient's own cells and avoiding embryonic cell use, iPS cells still have many problems. Transfection of genes has inherent difficulties. Viruses are typically used to delivery genetic material to generate iPS cells since they have natural mechanisms that can penetrate the cell wall and release nucleotides into the tight security of the cell nucleus. The downside is that viral gene delivery by retroviral and even lentiviral vectors causes random insertion into the host genome and could lead to unwanted mutations [194]. Adenovirus vectors, as well as non-viral

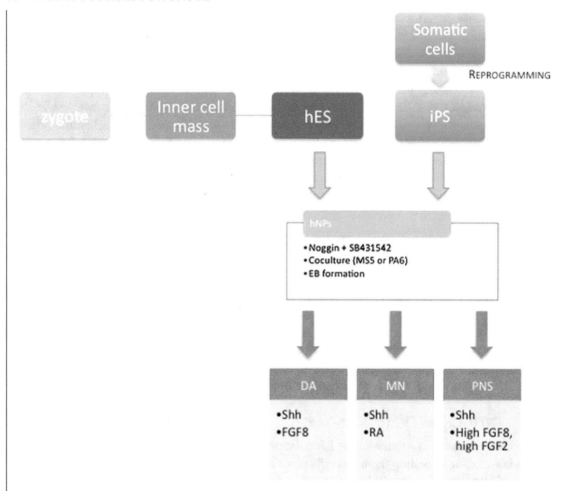

Figure 4.3: Diagram of differentiation paths for ES and iPS cells into specific populations of neurons. hESCs, human embryonic stem cells; hips, human induced pluripotent stem cells; hNPs, human neural precursor cells; DA, dopaminergic neurons; MN motor neurons; PNS, peripheral nervous system derivatives; EB, embryoid body. Image reprinted from [201].

methods, do not carry the same threat of random genome insertion but are often less efficient [194]. Alternative methods to viral gene delivery have been developed that are more effective, involving nanoparticle delivery or the use of commercial transfection kits [195–197]. Electroporation using commercially available systems allows increased permeability of the cell and nuclear membrane and has been used to deliver a single plasmid containing all four genes for induction of iPS cells [195]. Other commonly used methods include complexation with a cationic polymer or lipid to stabilize DNA and cross the cell membrane [198]. In depth discussion of other nonviral systems and subsequent uptake into the cell and nucleus can be found in other reviews [199, 200].

Although iPS cells are still far from clinical studies in CNS TE, they are being actively developed for diagnostic uses. It is proposed that patient specific cells could be used to screen drugs and therapeutic treatments, or conversely to study genetic and contracted diseases using patient iPS derived neurons or glia [201]. Disease specific iPS cells are currently being used to investigate neurodegenerative disorders including ALS, Parkinson's, and Huntington's disease, among others [202–204]. Continued work with iPS cells may prove them valuable in CNS TE approaches in the future.

4.3   ADULT STEM CELLS

4.3.1   ENDOGENOUS STEM CELLS IN THE BRAIN AND SPINAL CORD

Stem cells niches throughout the postnatal body retain populations of multipotent cells into adulthood (Fig. 4.4). In contrast to ES or iPS cells, NSCs within the brain and spinal cord closely interact with their surrounding cells and are committed to the neural lineage. Adult NSCs may also offer a more popularly acceptable source of stem cells because they lack the stigma that surrounds embryonic research and possess decreased tumorigenicity. Defined populations of multipotent adult NSCs have been identified in the subventricular zone (SVZ) (Fig. 4.4B) of the lateral ventricles, the dentate gyrus (DG) of the hippocampus, and around the central canal of the spinal cord (Table 4.1) [205].

SVZ and DG cells can differentiate into neurons, oligodendrocytes, or astrocytes [206, 207]. *In vitro*, these two stem cell populations behave very similarly. Both areas in the brain produce neurons under non-pathological conditions. Physiological neurogenesis in the DG takes place in the innermost region of the subgranular zone of the hippocampus, quickly becoming neurons with mature phenotype that incorporate into the granule layer [208–211]. Cells from the SVZ naturally migrate along a track lined by astrocytes to the olfactory bulb where they differentiate into granule cells [212–215]. The rostral migratory pathway is specific for these migratory cells, and transplantation of other neurons into the SVZ does not result in migration to the olfactory bulb [212]. On the contrary, NSCs from the DG of the adult hippocampus can be transplanted into the SVZ migration pathway and will differentiate into neurons typical of the olfactory bulb [216]. Adult NSCs are of interest for endogenous activation or exogenous transplantation for TBI or brain diseases owing to their multipotency and natural brain origin. Their implantation has been shown to decrease recovery time in SCI and enhance tissue bridging when used in conjunction with a nerve guidance conduit (NGC) (Fig. 4.5) [217].

Due to the delicacy of brain surgery and implantation, endogenous activation of the SVZ or DG populations is a motivating interest for many researchers. Multiple injury models have been used to stimulate cell activity in the NSC niches of the brain. In a study by Rice et al., TBI was induced in adult rats and proliferating cells were labeled and counted in tissue sections, which revealed increased proliferation and migration of cells from both the SVZ and DG post injury [218]. Interestingly, results from this study showed significantly timed 'waves' of increased mitotic activity

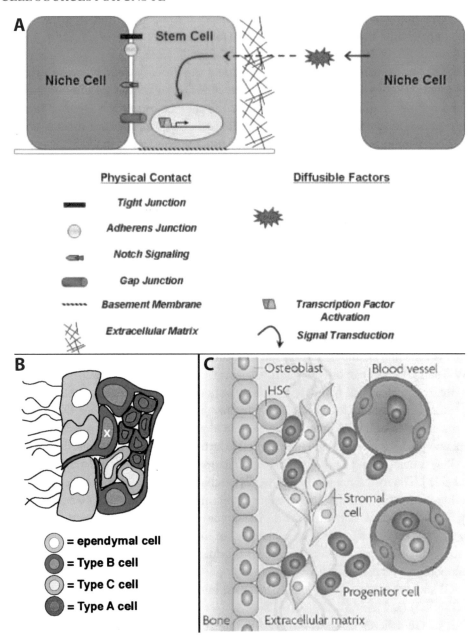

Figure 4.4: (A) General stem cell niche illustrating the multitude of cues that the niche environment encompasses. (B) SVZ niche containing astrocytes (type B cells) in close proximity with neuroblasts (type A cells). (C) Bone marrow niche. Images reprinted from [495–497].

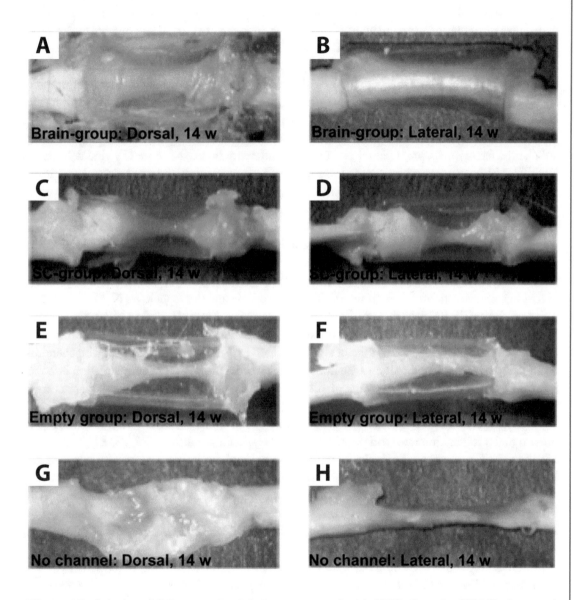

Figure 4.5: Spinal cord full transection injuries were treated with NSCs from the SVZ (Brain-group) or the spinal cord (SC-group) coated inside chitosan nerve guidance conduits. Analysis at 14 d showed improved tissue bridging in chitosan tubes containing NSCs from the brain (A, B) and the spinal cord (C, D) over acellular chitosan tubes (E, F) and the control (G, H). Image reprinted from [217].

from the DG. Similarly, after traumatic axonal injury, proliferation and migration of SVZ and DG cells, as well as an increased gliogenesis, persisted even eight weeks post-injury [219]. After inducing demyelinated lesions in the mouse brain, SVZ cells were observed to expand and migrate towards the lesion [220]. Observations show that NSCs in the adult brain respond differently to different injury models, presenting astrocytic preference in response to TBI and oligodendrocyte preference after demyelination [221]. Alternatively to injury-induced activation, injections of different neurotrophic agents have been studied for stimulation of growth, migration, and differentiation of NSCs in the brain. System perfusion of insulin-like growth factor I (IGF-I) in rats produced marked proliferation and neuronal differentiation in the DG without any increase in astrocyte production [222]. In the rat SVZ and olfactory bulb, proliferation and neuronal differentiation was induced by viral delivery of BDNF [223]. New neurons persisted 5-8 wks after the adenoviral injection into the lateral ventricles. It is clear that NSCs in the brain will respond to injury and a local or systemic delivery of neurotrophic agents. A properly designed CNS TE system has the potential to augment the stimulation of large populations of new neurons and glia from the SVZ and DG.

In vitro, SVZ derived adult NSCs can be induced to differentiate into neurons and glia by a variety of factors including IGF-I, NGF, BDNF, angiopoietin-1, cAMP, BMP-2, platelet derived growth factor (PDGF-AA), and the cytokine interferon gamma (IFN-γ), among others [224–228]. IGF-I has similar action *in vitro* to *in vivo*, and is mitogenic and neurogenic on NSCs [229, 230]. The combinatorial administration of IFN-γ in combination with an analog of cAMP to SVZ NSCs leads to a significantly higher population of neurons than soluble administration of neurotrophins, NGF, and BDNF, or tumor necrosis factor alpha (TNF-α) [227, 228, 231]. Even though biochemical treatments have been found to induce neuronal and glial differentiation for transplantation into the injured CNS, the question still remains whether these cells will integrate into the existing network of tissue, resulting in functional recovery.

Ependymal cells located near the central canal of the spinal cord have been identified that possess high proliferation rates and multipotency under specific conditions [22, 23, 205, 232–235]. Similar to NSC responses in the brain, injury of the spinal cord results in a protective response of the spinal cord ependymal cells [232, 233, 236]. Additionally, introduction of EGF and bFGF into the spinal column leads to increased proliferation of ependymal cells [237]. These results are expected, since EGF and bFGF are mitogenic agents, well known for their use in the *in vitro* expansion of NSCs. Ependymal NSCs were less effective than SVZ NSCs as a direct cell treatment for SCI in rats (Fig. 4.5) [217].

Obtaining adult NSCs from the brain and spinal cord is difficult and risky to the patient, which is the biggest drawback of these cells. However, isolating these cells has been shown possible and further advances in surgical techniques will increase their possibility for clinical use [238]. Another drawback of adult NSCs, in contrast to pluripotent cells such as iPS and ES cells, is that these cells show decreased proliferation with age [239]. Research with NSCs is still extremely useful as it could be applied directly to NSCs derived from ES or iPS cells. TE strategies that target endogenous activation of stem cell populations in the brain and spinal cord would be useful cues to include in

CNS constructs. In this respect, incorporation of these signals into a degradable scaffold could allow for a long release time to facilitate long-term regeneration by native tissue stimulation.

4.3.2   MESENCHYMAL STEM CELLS

MSCs have also been utilized for the purpose of CNS therapies. These multipotent stem cells are derived from a number of locations in the adult body and have the capability to differentiate into many different types of tissues with varying efficiency [240]. One of the main sources of MSCs is the bone marrow (Fig. 4.4C); these cells can be differentiated into smooth muscle cells, osteoblasts, chondrocytes, cardiomyocytes, liver cells, SCs, and to some degree, neurons [241–246]. In addition, MSCs isolated from the umbilical cord have similar characteristics and can express neuronal phenotypes following neural induction [247–249].

MSCs are able to promote regeneration in neural TE; however, experts are unsure of the therapeutic mechanism by which they accomplish this. MSCs may transdifferentiate into neurons and glia to augment tissue regeneration; alternately, inflammatory and immunological agents may be recruited to the damaged site by way of MSCs [250–252]. A secondary protection mode of MSCs is their secretion of neuroprotective factors that provide neural sparing and encourage endogenous axon regeneration. Crigler *et al.* studied MSCs for gene coding and expression of neurotrophic factors and found cultured MSCs could express BDNF and NGF [253]. Further testing showed that neuronal blastoma cells as well as explanted DRG both co-cultured with MSCs demonstrated significantly increased neurogenesis and neurite outgrowth. In these studies, inhibition of BDNF activity resulted in small decreases in outgrowth and proliferation, but the effects of MSCs were not completely negated, suggesting that other neurogenic factors are also secreted by MSCs.

Transplantation of undifferentiated MSCs into SCIs has shown positive therapeutic effects. Adult bone marrow MSCs transplanted two days after a spinal contusion injury showed neuro-protective effects that led to increased myelination at the injury site over the control group [254]. Histological analysis revealed increased laminin expression with MSCs along with neurite extension and alignment with spinal cord direction, but functional analysis using BBB scoring showed no significant difference between MSC treated injuries and the controls. Results suggested that MSC expression of laminin was responsible for aiding neurite alignment. As mentioned above, MSCs are also known to differentiate into myelinating cells. Three days after a focal demyelinated lesion was made in adult rats, undifferentiated MSCs delivered to the site helped to improve electrical conduction velocity across the lesion [255]. Spinal cord axons treated with MSCs contained myelination typical of the PNS, suggesting MSC differentiation and myelination within the wound. Direct injection and intravenous administration of MSCs in response to a demyelinated lesion in the rat spinal cord revealed that remyelination occurred in a dose-dependent fashion, as shown by histological examination [256]. This work also demonstrated the effectiveness of intravenous delivery of MSCs; the treatment is less invasive, although, almost 100 times more cells had to be used. MSCs also have aided healing in brain injury models. In an *in situ* environment, co-transplantation of MSCs with NSCs into hippocampal slice cultures led to the majority of NSCs differentiating into oligodendrocytes [257]. In contrast, when the NSCs were transplanted in the tissue slices alone, the majority became astrocytes. This effect translates to *in vivo* studies as well. A recent study involving MRI tracking of tissue and cortical blood flow (CBF) showed that administration of MSCs in the brain significantly reduced ventricle expansion and helped maintain CBF in regions adjacent to injury [258]. Compared to saline injections, the MSC-treated rats demonstrated significant functional improvements as assessed by neurological severity score and the Morris water maze test.

From a TE standpoint, MSCs could provide an efficient and elegant way of administering therapeutic agents in multi-cued constructs for CNS injury and disease. They provide neuroprotection, can be manipulated to secrete many growth factors, and can differentiate into myelinating cells. Ongoing research suggests MSCs could be used as a source of growth factors as well as for inflammatory and immune agents. Thus, incorporation of MSCs may address several issues at once. Isolation of cells from a non-invasive, although painful, bone marrow biopsy procedure is also very attractive to limit extra surgical procedures outside of treatment. In depth investigation of the long term effects of MSC transplantation are underway, and based on the above findings, MSCs present a promising choice for cell therapy for CNS regeneration applications.

4.3.3   NEURAL CREST-LIKE STEM CELLS

Skin-derived progenitor cells (SKPs) are a population of cells within the skin that have neural-crest cell capabilities [259–263]. These multipotent SKPs were first described in 2001, and have since been shown to possess high proliferative ability and the capability to differentiate into neuronal and glial lineages [261, 262]. During development, the neural tube forms from the ectodermal germ layer and neural crest cells migrate from here to become a variety of cell types, including peripheral neurons, SCs, melanocytes, smooth muscle cells, connective tissue cells, cartilage, etc. [181–183]. SKPs can be cultured similar to adult NSCs in substrate free media with the mitogenic agents EGF and bFGF to produce neurospheres [261, 262]. SKPs have been shown to express nestin, as discussed earlier, a marker for NSCs [264–266].

The transplantation of SKP stem cells into SCIs has afforded functional recovery in walking tests; however, in the study SKPs were not compared to any other cellular treatments (Fig. 4.6) [267]. Neuronal differentiation of SKPs has had limited success so far, and very little success in serum free media [265]. Low concentrations of serum combined with neurotrophins, results in β-III tubulin positive cells co-expressing neurofilaments; however, after patch-clamping analysis, the cells were found to have weak electrophysiological profiles [265]. The development of a more robust differentiation protocol is required to yield significant populations of functioning neurons from SKPs. To date, better success has been achieved in directing SKP differentiation to a glial-lineage. Over 80% pure populations of SCs from SKPs are possible utilizing efficient protocols with the differentiation cue Neuregulin-1β [268, 269]. As mentioned previously, SCs have been shown to remyelinate CNS axons in the spinal cord even though they are typically found in the PNS. In fact, SKP derived SCs were shown to have excellent morphology and myelinating capabilities,

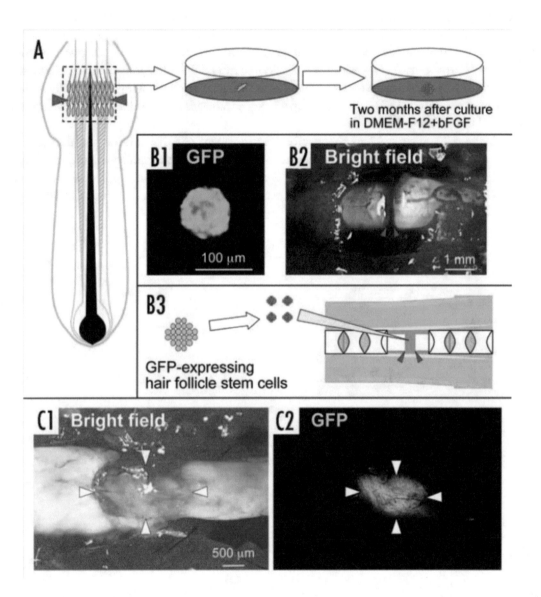

Figure 4.6: (A) Schematic representation of SKPs being isolated from the hair follicle bulge and cultured in neurosphere forming media. (B1) Example of GFP staining in a SKP neurosphere. (B2) Transection injury in the thoracic region of 6-8 week old mouse. (B3) Schematic representation of transplantation into the transected spinal cord. Two months post-transplantation the SKPs have rejoined the spinal cord, shown in brightfield (C1) and GFP (C2). Image reprinted from [267].

significantly higher than undifferentiated SKPs, when transplanted into a contusion injury in the rat spinal cord [268].

One flaw that is redundant throughout adult cell therapy techniques is the decreased plasticity and activity of aged cells. A comparison of SKPs derived from different regions of skin from patients 8 months to 85 years old revealed that proliferation and differentiation capabilities drastically declined in cells from the elderly [270]. SKPs still have many advantages and continue to be developed for additional CNS TE applications.

Many cell sources exist to choose from for treatment of CNS injuries and diseases. The specific needs of TBI, SCI, and neurological diseases call for tailored approaches when developing and optimizing neural regenerative methods. TE constructs that incorporate cells and/or encourage ingrowth have the capability to augment and accelerate the restoration of normal CNS function, improving quality of life.

CHAPTER 5

Stimulation and Guidance

Multi-cued CNS TE constructs are made possible through incorporation of surface and soluble prompts combined with scaffolding (Chapter 3) and cell components (Chapter 4). Cell migration in combination with process extension are essential early steps in nervous system development; thus, developmental signals are often the source for stimulation schemes aimed at controlling cell behavior and reinnervation after injury Immature cells are partitioned and guided to the appropriate targets by a complex mixture of chemical and physical cues. During nervous system development, neurons rely on the surrounding environment for guidance in order to innervate their targets and create a functional neural network system. Initially, neurites extend from the soma in all directions, and some gradually unite to become the primary axon while remaining neurites form dendrites [271, 272]. The growth cone, located at the tip of the axon (Fig. 2.2A), serves to decipher chemical and physical cues, both in space and time, determining the axon's trajectory [273, 274]. As it senses these signals, the growth cone pauses and enlarges as the cytoskeleton reorganizes, preparing itself for the next move [275]. As highlighted in Chapter 2, in the periphery of the growth cone are located lamellipodia that contain a web of actin. From this region, finger-like projections called filopodia protrude and sense the environment. Microtubules are responsible for transmitting this information from the growth cone to the soma and back to the tip. Axon branching can occur where the growth cone will split or interstitial branches will form from actin remnants as a result of cytoskeletal changes [275]. Guidance of the growth cone is vital to properly guide axons to their targets. Several strategies for encouraging and directing growth cones to specific targets will be introduced including physical, chemical, and electrical applications. Most can be used in conjunction with each other for complex control over neuronal behavior.

5.1 PHYSICAL CUES

Guidance and stimulation of cells can result from physical connection with the environment around them. Cells are sensitive to changes in substrate stiffness, mechanical stresses as well as topographical changes. Analyzing how physical cues modulate cellular adhesion, differentiation and guidance will provide a framework for developing optimal biomaterials and bioreactors to optimize translation of neural TE constructs *in vivo* (Fig. 1.1). Therefore, material stiffness, physical elongation and topographical guidance methods used for neural regenerative strategies will be outlined in addition to a brief review of topographical fabrication techniques.

5.1.1   PHYSICAL STIMULATION

Material Stiffness

As mentioned in Chapter 3, many biomaterials used in TE have tunable mechanical properties that can be adjusted by material blending and degree of crosslinking. This is important for biological applications because cells tend to favor their native tissue moduli. Additionally, during development and in certain disorders, changes as well as gradients in ECM stiffness are important for proper development and healing responses. Cell interpretation of mechanical stimuli and the resulting response, including the activation of downstream pathways, is known as mechanotransduction. Vascular tissue and cell migration mechanotransduction is better understood compared to responses in the nervous system; thus, there are still many questions concerning the pathways activated in response to mechanical forces in the nervous system [276]. Cells adhere to surfaces and exert contractile forces in order to migrate to different areas. In CNS and PNS TE, low substrate elastic modulus (E) is important for neuronal outgrowth [277, 278]. Several studies using DRG have shown that gels with lower stiffness result in more branching and longer neurites than stiffer gels [279–281]. In a study by Flanagan *et al.*, mouse spinal cord neurons and glia were grown on polyacrylamide gels with elastic moduli ranging from 50 to 550 Pa [277]. This group found that on the softer acrylamide, neuron branching was three times higher and glial survival decreased. Gradient stiffness gels can induce neuronal durotaxis, as neurites have grown significantly longer down a decreasing stiffness gradient [280]. Interestingly, a threshold effect has also been observed when stiffnesses were higher than a shear modulus (G) of 100 Pa (E $\cong$ 600 Pa); neurite outgrowth was still present at higher moduli but was not as pronounced as in the range of G = 10-100 Pa (E $\cong$ 60-600 Pa) [282]. Differentiation into a specific lineage or cell type can be specified by culture substrate compliance. Leipzig *et al.* found that extremely soft gels with stiffness, E around 800 Pa, will elicit neuronal differentiation from NSCs, while slightly stiffer gels with an elastic modulus around 7,000 Pa yields an oligodendrocyte phenotype [94].

Physical Elongation

The neuronal growth cone is indeed an important way that axons extend towards a target, but it is by no means the only method of elongation in the axon. During development, growth of the organism continues after synapses are formed. To maintain the neuronal network, the axon must continue to grow in response to the continued tension placed on them. This process has been exploited in a number of experimental settings to elongate axons to great distances [12, 283–285].

Axons can be elongated by micropipette towing of the terminus [286–288]. This pioneering method of axon stretching has been used to investigate mechanical properties of axons. Alternatively, the Smith lab has been very active in extreme axon elongation [12, 283–285, 289–291]. They have developed a device that can stretch millions of axons at a time and accelerate axonal stretch rates up to 1 cm/d without breakage or thinning. Embryonic rat DRG cells were seeded across two substrates to allow neural cell bodies to adhere to each; the substrates are then separated in a step-rest pattern causing the axons between the two populations of cell bodies to be elongated. Stretched constructs have been shown to retain the same electrophysiological competence as control cultured neurons by displaying similar voltage channel density and patch clamping [289]. When stretch-generated constructs were implanted into rat sciatic nerve lesions, they showed promising integration into host tissue and axons within the tubular graft displayed signs of host myelination [292]. The promise of tension-induced axonal elongation for use in spinal injuries is enticing. Damaged tissue could be excised and bridged by pre-grown neurons stretched to the correct length, cutting down on healing time since large gaps would already be filled with existing axons. Iwata *et al.* teamed up with Smith to create stretch grown constructs and to test their ability to repair a rat hemisection SCI [293]. Embryonic DRG were elongated to 10 mm using tensile elongation, encapsulated in collagen and implanted 10 days after injury. Tissue bridging was observed after four weeks, but the functional benefits were not significant over collagen alone. Thus, proper interfacing and reconnection with host tissue is still a concern, especially in the spinal cord where there are complex organizations of axon tracts.

5.1.2   PHYSICAL GUIDANCE

Physical cues can be used to directionally stimulate cells for guidance strategies. Advancements in microfabrication techniques have allowed for new methods of surface patterning to be generated with enhanced resolution. A few common surface manufacturing approaches will be discussed briefly followed by specific patterns attempted both *in vitro* and *in vivo* for neural guidance (Fig. 5.1). Although the specific cellular effects and changes in signaling due to topography have yet to be elucidated, some theories on cytoskeletal and functional mechanisms will be mentioned. Further review of topographical and surface guidance can be found in the following papers [294–296].

Fabrication Methods

Many of the microfabrication techniques used to make topographical surfaces for cell behavioral control utilize lithographic methods. Very often during the fabrication of neural culture surfaces, soft lithography is used to transfer a substance or molecule from the raised pattern of a soft rubber mold to another substrate. The molecule can be a polymer, bioactive factor, or a chemically reactive compound for further conjugation. Microcontact printing is a form of soft lithography that typically uses a polydimethylsiloxane (PDMS) stamp to transfer a self-assembled monolayer onto a substrate. Another lithographical technique, which is often used to make the original template for soft lithography, is photolithography. This technique is similar to chemical etching in many ways. A photomask is used to selectively protect and expose specific regions of a surface to light. A photoresist material coating the desired surface is polymerized in the lit regions beneath the photomask. After light exposure, any unpolymerized photoresist from masked regions is removed. From here, a PDMS mold can be cured on the pattern, or further processing can be performed to use the original substrate for cell culture (including deposition of a coating or polymer).

The production of biomimetic substrates has become increasingly popular for topographical cell response studies. Research has demonstrated that during neural development neurons can often

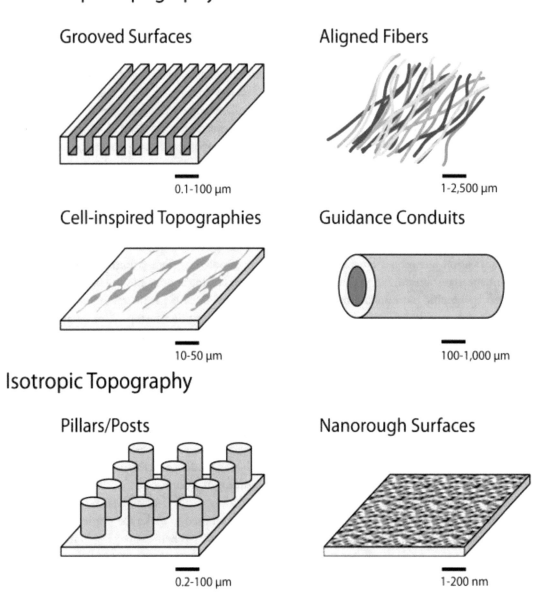

Figure 5.1: Illustration of topographies that can be used in cell guidance, on the nano or micro scale. Image reprinted from [294].

be guided by direct interaction with glia. In response, patterning of live cells is being developed in an attempt to create guidance strategies for neural TE. Recently, a cell printing method has been developed (similar to an inkjet printer for paper) to deposit bioadhesive factors or live cells themselves [297–302]. Alternately, lithography methods have been used to pattern or grow a cell substrate with particular alignment to subsequently grow neuronal cells on the surface [303–305].

Electrospinning, as covered in Chapter 3, is a method that allows imitation of portions of the ECM nanoenvironment. To briefly recap, a polymer is dissolved into a solvent prior to electrospinning, and pumped through a needle through an electric field. The fiber travels across the electric field to a substrate that can be stationary or in motion to allow for specific fiber orientation. Random and aligned fiber meshes can be created with electrospinning with fiber diameters ranging from nano to micro scale. Polymer concentration, flow rate, electric field strength, collection gap, and even collector speed can be adjusted to tune the fiber size [103, 104]. Cell encapsulated electrospinning has been developed using coaxial needles (cell core with polymer shell), but to date this has not been applied to neural cells [306, 307].

Cellular Response to Topographical Designs

One of the most popular anisotropic patterns for cellular alignment is grooved or ridged substrates. These can be made in a large range of feature sizes by adjusting the height, width, and spacing of the grooves in the nano and micro scale. Micron scale grooves have been used with different cell types to induce specific guidance, but with mixed results. With neurons, smaller grooves (<10-20 μm) tend to cause a higher occurrence of perpendicular alignment (across grooves), but also include parallel alignment of neurites to grooves [308–310]. Recently, a phenomenon of cell bridging across micron sized grooves was observed with DRG neurons, hippocampal neurons, SCs, and neuroblastoma cells (B104) [311]. Each cell type was seen to extend processes across adjacent plateaus (spaced 30-100 μm), especially with increasing cell number and plateau width (30-100 μm width). In addition to directional orientation, microgrooves have been shown to induce neuronal polarization (axon establishment). Embryonic hippocampal neurons on 1 to 2 μm wide grooves were shown to have significantly higher occurrence of polarization than on flat PDMS, especially when groove depth was increased from 400 μm to 800 μm [308]. From a TE standpoint, this work could be useful in inducing polarization of differentiating stem cells for CNS work. It is also important to note that the current trend is toward the development of degradable, as well as, surface patterned materials. Traditionally, most topographical studies have been performed on PDMS or other non-degradable materials due to their ease of fabrication with current techniques. In order to transition to implantable materials that are biodegradable (for reasons discussed in Chapter 3), some researchers have begun utilizing PLA for grooved substrates [310, 312].

Another common physical pattern, especially for neural TE, is channels. Large tubular NGCs (nerve guidance conduits) and smaller channels imitate white matter tracts found in the CNS and are therefore often used to lead spinal cord axons parallel to the spinal column during regeneration (Figs. 3.3 and 4.5). NGCs are the current preferred bioengineering strategy for regenerating the

PNS. Yu and Shoichet synthesized longitudinally oriented multichannel NGC that increased DRG adhesion and outgrowth especially in peptide modified channels [313]. However, NGC applications can be utilized in the CNS, as mentioned previously in the materials section, where agarose channels were implanted into a rat spinal cord (Fig. 3.3) [116]. Also, NGCs incorporating additional components such as neurotrophins or supporting cells were able to generate enhanced axon growth, suggesting that NGCs could be useful for SCIs [314, 315]. Recently, one study synthesized semiconductor nanotubes (1 to 10 μm in diameter) out of silicon and germanium and formed them into arrays that encourages single axon penetration and outgrowth into the nanostructure [316]. This model electrically and physically mimicked the native myelin and therefore could be utilized for future neural network applications.

The ECM-like morphology of fibers with small geometries has been used to stimulate and guide cells. At a cell and structural level, the ECM ranges in size from 50 to 500 nm [317, 318]. Nanofibrous scaffolds have been created using three main techniques: self-assembly, phase separation, and electrospinning (see review [319]). Aligned fibers have been used in neural cell culture to induce neurite sprouting and guidance [320, 321]. Fiber diameter is also an important factor to consider in optimizing neural outgrowth and guidance [322, 323]. Although not a true fiber scaffold, the nano surface topography of nerve basal lamina was recently replicated using a PDMS stamp on a polystyrene substrate [324]. The resulting topography resembled nanofibrous laminin structures, and supported the growth and alignment of DRG neurons. This study illustrates that cell reaction to fibrous scaffolds is very much tied to the surface structure and the feature size must be small enough for cells to sense and interact with it.

As a substitute to using actual cell topographies in co-culture, biomimetic surfaces aimed at reproducing these structural cues have been fabricated. Microcontact printing has been used to align SCs, which in turn were used for neural guidance [325, 326]. In a creative study by Kofron *et al.*, PDMS was used to make both aligned cell topography (taken from live cultures), and mimetic cell topography from a computer assisted design (CAD) program [303]. The CAD generated topographies were tested against live cultures of astrocytes, endothelial cells, and SCs for DRG neurite alignment and length and displayed comparable or improved results in the CAD topographies as compared to the live topographies. Kofron and Hoffman-Kim also optimized and quantitatively analyzed cellular monolayers of astrocytes, endothelial cells and SCs using Design of Experiment (DOS) and Response Surface Methodology (RSM) [327]. The statistical experimental design of this study afforded insights into the micropatterned feature sizes that affect cellular adhesion, alignment and confluence. These powerful tools, along with other modeling programs, take into account the multiple and complex biological variables and are meant to alleviate time and resources spent on bench top experiments.

5.2   CHEMICAL CUES

Several biochemical guidance signals have been identified including: chemoattractive factors such as neurotrophins and netrins, chemorepulsive agents like semaphorins and slits, or contact-mediated molecules such as ephrins and those located in the ECM (Fig. 5.2). Deciphering biomolecular guidance activity during nervous system development as well as injury is key to generating new techniques and tactics for improving and restoring function to the nervous system after injury. For this reason, general immobilization techniques and specific PNS and CNS chemical guidance strategies will be preceded by an overview of promising molecules commonly used in these studies.

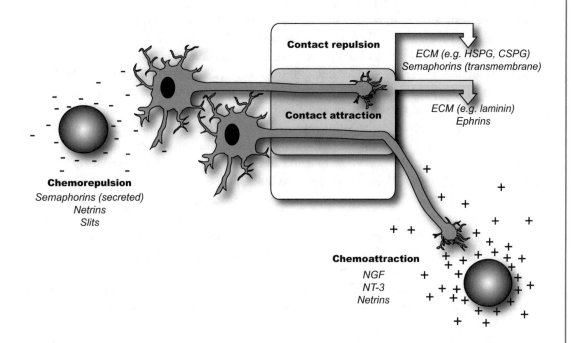

Figure 5.2: The extracellular matrix (ECM) alongside a combination of contact-mediated and soluble factors guide the axons of maturing neurons to their innervating target during development. While the attractive cues pull on axons, the repulsive cues push axons to the path towards their proper targets. Images reprinted from [6].

5.2.1 EXTRACELLULAR MATRIX

Proteins and polysaccharides primarily make up the complex structural framework of the ECM, which is secreted and organized into specific tissues by cellular architects. In turn, the assembled ECM plays a vital role in regulating cell behavior throughout development and adulthood, providing anchorage, mechanical buffering, segregating different tissues and aiding cell-cell communication.

The ECM is paramount to progenitor cell migration and differentiation, as well as to axonal genesis and extension.

Throughout the body, laminins serve as a key component of the ECM, providing binding sites for self polymerization, other ECM macromolecules and cells. They are key contact mediated cell adhesion promoters for neural cells. To date, fifteen laminins have been characterized and all share a similar trimer structure that is generated from five α, four β and three γ chains [328]. Within the nervous system, laminins make up basement membranes that are essential to interactions between neurons and glia. Research has demonstrated that laminin is required for neuronal migration in the developing cerebellum [329, 330]. Laminin contains several binding motifs that interact predominantly with cell-surface integrins (primarily $\alpha 1\beta 1$ and $\alpha 6\beta 1$) and secondarily with α-distroglycan [331, 332]. Interactions between integrin receptors to their ligands contained in ECM proteins are central for anchoring and directing the growth cone in order to initiate guidance [333, 334]. More detailed information regarding integrin interaction and signaling is reviewed elsewhere [335–337]. Jacques *et al.* [338] have showed that neural precursor migration on laminin can be significantly halted by inhibition of $\alpha 6\beta 1$ integrin subunits. Developmental studies have observed that knocking-out $\beta 1$ integrin expression blocks basement membrane formation and the expression of laminin protein [339]. Laminin provides essential attachment points enabling axons to extend and exert forces on the ECM [340–342]. Collagens and fibronectin are also important in the nervous system and are generally supportive to neuronal cell migration and neurite extension [343, 344]. Fibronectin, laminins, and collagens activate similar cell integrin receptors; thus, all three are important in cell substrate adhesion.

Proteoglycans are important ECM molecules that have been implicated in neuronal and axonal guidance. To review their introduction in Chapter 2 , these highly negatively charged molecules consist of a core protein with numerous GAG side chains and are grouped into two major classes: HSPGs (heparan sulfate) and CSPGs (chondroitin sulfate). Several studies have demonstrated that either exogenous addition of HSPG or enzymatic removal of HSPG leads to axonal guidance defects during development [345–347]. This has led to findings that demonstrate a functional association between HSPGs and several secreted and transmembrane proteins. As a result, the role of HSPGs in guidance appears to be tied to sequestering slit, netrin and semaphorin proteins (thoroughly reviewed in [348]). The influence of CSPG on guidance is not as well understood as that of HSPG; however, it is clear that CSPG also has a potent inhibitory effect on neuron guidance. Studies have demonstrated that CSPG makes up part of the glial scar that effectively halts CNS axonal regeneration, and leads to the generation of abnormal axonal growth cones [349]. Work has shown that, similar to HSPGs, CSPGs also functionally associate with semaphorins and inhibit axonal extension [350]. Li *et al.* utilized microfluidic techniques to create parallel and opposing gradients of laminin and CSPG [351]. They found that cultured DRG neurons show preference for higher laminin and lower CSPG in opposing gradients by exhibiting strong axon directionality.

Tenascins are ECM glycoproteins primarily involved in development as well as wound healing. The tenascin (TN) subtypes, TN-C and TN-R, are two of the four TN family members found in the nervous system and are generally repulsive to axons. However, TN-C and TN-R contain multiple domains that can be either repulsive or attractive depending on presentation, and may provide more precise control of neurite extension, migration and guidance [352]. The inhibitory effects of TN-R on neurite outgrowth can be overcome by laminin and fibronectin used in combination, as shown in a RGC outgrowth assay [353].

5.2.2   NEURAL GUIDANCE MOLECULES

Besides the neurotrophins, the molecules and receptors described below have largely been studied for their guidance and signaling importance in development and injury. A better understanding of these molecules and their interactions will help in the future formulation of approaches for enhanced control of axonal guidance, branching, pruning and synapse formation for the purposes of regenerative medicine and TE. The following discussion is by no means exhaustive, especially in regard to developmental findings and signaling pathways. For more complete reviews please see [14, 354–356].

Neurotrophins

The neurotrophin family is composed of secreted chemoattractant proteins that are integral to nervous system development and maintenance [357]. NGF, NT-3, BDNF, and neurotrophin-4/5 (NT-4/5) are the four major neurotrophins and are similar in structure and sequence [358]. Tropomyosin receptor kinases (Trks) and p75 neurotrophin receptors (p75NTRs), located primarily on the terminal end of axons, are vital for the initiation of neurotrophin signaling [357, 359]. Three Trk receptors have been identified to date. NGF binds specifically to TrkA [360–362], whereas, BDNF and NT-4/5 prefer TrkB receptors [363, 364]. NT-3 complexes with TrkC with high affinity but can also interact with the other Trk receptors [365, 366]. P75NTR shows high affinity for NGF and serves as a low-affinity receptor for all the neurotrophins [367].

The first neurotrophin to be discovered was NGF, which was shown to encourage neurite extension from sensory ganglia *in vitro* [359]. Subsequent work found that NGF exists at significant concentrations in adult tissue showing neuronal specificity; however, NGF has been shown to also elicit responses from other tissues and cell types [359]. The existence of other neurotrophic factors was postulated early on since neurons depend on specific targeting for proper innervation during development. BDNF was the next neurotrophin identified from mammalian brain isolates and has been shown to significantly act on CNS and directly associated neuronal populations [368, 369]. BDNF has been shown to play an important role in the regulation of synapse structure and function, especially in glutamatergic synapses [370]. NT-3 shares significant homology to NGF and BDNF; however, NT-3's primary role occurs during the development of many tissues as indicated by the broad distribution of its messenger RNA [371–373]. Both BDNF and NT-3 have been shown to significantly influence axon path-finding, as well as aiding axonal regeneration in rats following SCI [374, 375]. Moreover, during development both neurotrophins direct the path-finding of maturing axons to their targets by a combination of long range attractive and repulsive cues [376].

NT-4/5, also known as NT-4 and NT-5, influences the survival and outgrowth of sensory and sympathetic neurons [363, 377, 378]. The importance of these neurotrophins in the nervous system have been reviewed in [379]. New experimental evidence has determined that bFGF and GDNF may also be important for nerve regeneration in the PNS and CNS. Both growth factors have been shown to influence neurons, SCs, and oligodendrocytes towards axonal growth and remyelination following injury [380, 381]. Finally, CNTF has been shown to act primarily on neurons as a survival factor following injury in the PNS [382]. To date, CNTF has not demonstrated any functional or regenerative benefits for nerve repair [383]; however, it may show synergistic effects in the presence of other neurotrophins [384].

Ephrins

Ephrin receptor tyrosine kinases (RTKs) are a large family of membrane bound proteins that are largely known for their role in the regulation of axon guidance as well as cell migration, vascularization, tissue segregation and synaptic plasticity [14]. Ephrin ligands are classified based on their extracellular domain sequence, where A-ephrin subclasses are glycosylphosphatidylinositol (GPI) membrane anchored and B-ephrin subclasses are transmembrane proteins [385]. Even though they show preferential binding to their specific ephrin protein, some Eph receptors can be activated by ephrin ligands from the opposite subclass [386, 387]. Ephrin/Eph complexes are remarkable for their ability to transduce signals bidirectionally into the receptor (Eph) expressing cells as well as the ligand (ephrin) expressing cells in what is termed 'forward' and 'reverse' signaling [14]. Both directionalities have been shown to be important in developmental axon guidance. Work has shown that ephrin/Eph signaling can be modulated to control targeting. It was shown that ephrinA2-EphA3 binding in axons initiates protease cleavage of ephrinA2, resulting in axon detachment and termination of contact repulsion [388].

Ephrins play an important role in development as well as regeneration. EprinA2 expression has been shown to decrease after injury to the superior colliculus in the neonatal rat brain leading to aberrant projections of ephrinA2 sensitive RGCs and cortical-tectal axons [389]. It is believed that these new projections most likely result from the removal of ephrin's repulsive cues [390, 391] as well as the possible attractive signaling provided by cortical-tectal axons [390]. Artificial upregulation of EphA4 does not aid in regeneration after SCI; however, it may stabilize synapse formation by reducing deviating projections [392]. Recently, it was shown that EphB/ephrinB contact between fibroblasts and SCs in a wounded peripheral nerve bed activates the Sox2 signaling cascade in SCs, stimulating peripheral nerve regeneration [393, 394]. For a thorough review on ephrin/Eph interaction during development and neurogenesis please see [395].

Semaphorins

The Semaphorins are a large family of proteins that contain approximately 500 amino acids each and are broken down into eight subclasses based on characteristics such as membrane anchorage (transmembrane or GPI-linked) or secretion [396] with the vertebrate semaphorins making up subclasses

3-7 [397]. Semaphorins and their receptors have recently been reviewed in more detail in [398]. Semaphorins are largely known for their growth cone collapsing properties [399]; however, recent work has revealed sub-types that attract extending axons [400]. Neuropilin and plexin have been identified as the two primary semaphorin receptors that transduce semaphorin signals [401–404]. The majority of semaphorins utilize class specific domains to bind directly to plexin; however, class 3 semaphorins cannot directly bind to plexin [403, 404]. Instead, Sema3A binds to first to neuropilin, and this complex can then bind and activate plexin receptors. Plexin activation in turn activates the Rac1 (Rho GTPase) signaling cascade, leading to growth cone collapse and retraction [403, 404]. Class 3 semaphorins and their receptors are important in the development of many nervous system regions including the hippocampus, cortex, olfactory system as well as cranial and spinal nerves [405].

Semaphorins emit long or short range repulsive cues by presenting as either soluble or membrane bound molecules, respectively, and it is important to understand how they interact with their environment during injury to exploit their potential role in regeneration. Following adult sciatic injury, it was shown that the expression of five out of the six class 3 semaphorins (Sema3A, Sema3B, Sema3C, Sema3E, and Sema3F) as well as their receptors neuropilin-1 (NRP1) and neuropilin-2 (NRP2), were upregulated [406]. Class 3 semaphorins can act either in an attractive or repulsive manner following nerve injury; thus, at this time their exact role is not fully understood. For example, class 3 semaphorins were found to contribute to the inhibitory nature of the glial scar after SCI [407]. Conversely, NRP2 upregulation was directly associated with peripheral nerve regeneration following sciatic nerve injury in mice [408]. It was postulated that NRP2 signaling could lead to SC recruitment, which has been shown to offer vital cell support during axonal regeneration in the PNS [406, 408, 409]. Class 3 semaphorins dual nature has also been examined in other studies [410, 411] Semaphorins are typically not utilized for neural regenerative applications when attractive guidance biomolecules are more readily available; however, their repulsive and attractive signals could prove beneficial in optimizing TE strategies.

Slit and Roundabout

Slits are a class of secreted proteins that bind to roundabout (robo) transmembrane receptors. The ligand-receptor couple controls many processes in neuronal development including neuron migration, axon pathfinding, and axonal and dendritic branching [14]. Slit-1, slit-2 and slit-3 are expressed during development in the mammalian brain, spinal cord and thyroid, respectively [412]. Slit proteins act as a repellent cue to neurons during development [413, 414] and bind to four known receptors: robo-1-4 [415–417]. Interestingly, slit-2 is unique in that it can be proteolytically cleaved into a smaller, diffusible C-terminal fragment and a larger N-terminal fragment [418]. Each part of slit-2 has contrasting effects on different populations of cells; both the full slit-2 protein and the N-terminal fragment repel olfactory bulb neurons. Only N-terminal slit-2 fragments are able to induce DRG branching and extension [419]. The purpose of the C-terminal slit-2 fragment is not fully understood and cleavage of this end terminus may activate the N-terminal fragment for certain signaling pathways [419]. The robo-slit interaction plays a major role in commissural axon guidance at the midline during development [420, 421], assists in spatiotemporal patterning and neurite projections during brain development [422–424], as well as in properly directing axonal extensions in the optic system [425, 426].

Slit activity is vital to proper development, homeostasis and nerve regeneration. Slits and their receptors have been tied to CNS injuries. Slit-1 expression is upregulated in the cerebellum following injury, slit-3 upregulation occurs after SCI and microglial cells express both slit-1 and slit-3 in response to injuries in both regions [427]. Astrocytes have been shown to express slit-2 after brain injury in the hippocampus [428]. These findings suggest that slits and their repulsive nature contribute to regenerative failure after CNS injury. On the other hand, another study suggests that slit-1 supports axon elongation, whereas slit-2 is responsible for the remodeling of dendritic branches around the soma [429]. The HSPG, glypican-1 works in combination with slit-1 and robo-2 to directionally guide DRGs via repulsive cues as well as to prevent random, uncontrollable axonal elongation following both sciatic nerve injury and SCI [430]. This finding is surprising since these proteins were previously thought to only be present at development and not in both types of injuries. The existence of these molecules suggests that pruning and maintenance of the nervous system occurs after injury to the PNS and spinal cord.

Netrins

Netrins play a central role during spinal cord development, providing cues to guide commissural axons across the two halves of the spinal cord [431, 432]. They are a small family of secreted proteins with protein sequence homology to laminins. Netrins have complex interactions that can be attractive and repulsive depending on cell type and age of the host. Netrins interact with the transmembrane receptors neogenin and the receptor deleted in colorectal cancer (DCC); ligand receptor interactions initiate signaling pathways in commissural axons allowing for CNS interconnections [433–435]. Src family kinases and focal adhesion kinases act downstream of DCC and regulate growth cone adhesion and actin dynamics [436]. Since the subdomains of neogenin have similar homology to DCC, it is believed that netrin-neogenin binding activates similar pathways, leading to cell adhesion and axon guidance [437]. Uncoordinate-5 (UNC-5) is a family of netrin receptors that mediate axon repulsion by forming a receptor complex with DCC [438–440]. Netrin attraction can be silenced by slit-robo interactions via the formation of heteromultimeric receptor complexes [441]. Silencing of axons that cross the midline during development is extremely important, since these axons must quickly overcome their attraction or forever remain at the midline [14]. Following CNS injury, several netrin homologues play an important role in the regeneration of brain neurons and ventral nerve cord cells [442]. After rat RGC axotomy, DCC receptor expression is down-regulated, suggesting that if netrin-1 expression is targeted for nerve regeneration, its co-receptors must be targeted as well [443]. Therefore, it is important to consider the interrelationship netrins have with their receptors when modeling reinnervation techniques because they can be chemoattractive or chemorepulsive depending on what receptor is active.

5.2.3   TETHERING OR COVALENT IMMOBILIZATION OF NEURAL GUIDANCE FACTORS

Protein Immobilization Strategies

Protein patterning has gained significant attention with advances in techniques such as lithography, microfluidics, microcontact printing, and biosensors. Protein patterning has enabled high-throughput measurements of biological responses using micro and nano scale protein assays. There are three primary methods used to immobilize protein: physisorption, chemisorption and bioaffinity (Fig. 5.3). Physisorption is the physical adsorption of proteins to the surface via intermolecular

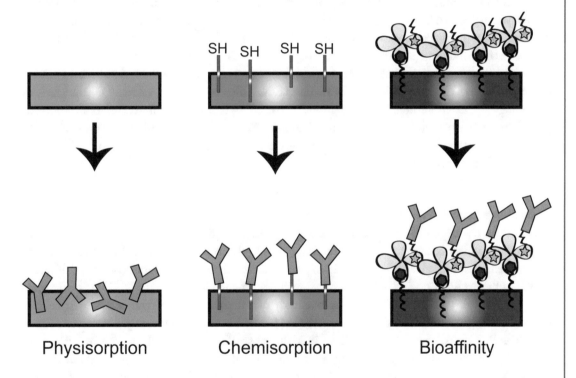

Figure 5.3: Common methods of protein surface attachment. Physisorption is the physical nonspecific adsorption of proteins to the surface. Chemisorption utilizes covalent bonding between protein functional groups and a modified substrate. This figure depicts thiol-maleimide interaction. Bioaffinity depicts the strong noncovalent affinity steptavidin (tetrameric protein) has with biotin (small molecule on the protein). This type of immobilization shows specific linkage of a protein to modified surface.

forces such as hydrophobic and polar interactions. This type of immobilization is nonspecific, such that proteins are adsorbed heterogeneously across the surface and are oriented to minimize repulsive forces with other molecules and the substrate. Adsorption is cheaper and easier than chemical means; however, the binding affinities are low and proteins can desorb easily, quickly leaving the original surface exposed. Chemisorption, on the other hand, is a type of immobilization that results in covalent bonding between exposed side-chain functional groups and modified surfaces. N-Hydroxysuccinimide (NHS)-amine, carboxyl, thiol-maleimide, epoxy, and photoactive chemistry are all common strategies used for covalent attachment. Often with these techniques, nonspecific chemisorption takes place as exterior residues containing the reactive groups on the protein attach to the surface. Nonspecific attachment could block active regions of the protein of interest or inhibit important conformational changes. Therefore, site-specific protein immobilization is desired to reduce unwanted attachments. This requires the insertion of functional moieties into proteins specifically for immobilization to surfaces with the corresponding coupling molecule. The biotin-avidin system is a well known biochemical immobilization strategy largely because it exhibits one of the strongest noncovalent bonds known ($K_d = 10^{15}$ M^{-1}) [444]. Streptavidin is a tetrameric protein that has a comparable affinity to biotin because it is structurally similar to avidin. Alternatively, histidine/nickel interaction, that allow tagged histidine regions on proteins to bind to nickel chelated complexes such as Ni-nitiloacetic acid (NTA), and DNA-directed systems, such as DNA microarray technology and DNA-protein bioconjugation, are two other types of bioaffinity immobilization approaches commonly used for protein immobilization [445–447]. These methods of immobilization are attractive due to the specificity and homogeneity of the oriented molecules. For more information on immobilization strategies, please see [448, 449].

Utilization of Immobilized Neural Biomolecules

Recent work has begun to reveal how tethering or immobilization of growth factors and guidance molecules modifies cell and stem cell function [450–459]. Immobilization of bioactive factors to biomaterial substrates allows for not only migrational stimulation, but spatial control of differentiation with sustained dosing, which is not possible with soluble factors [451, 455]. At the same time, cytokine immobilization allows for the study of the dynamics and the necessity of cellular internalization for activation of signal transduction [458]. Research has recently shown that IFN-γ as well as PDGF-AA can be immobilized to hydrogel scaffolds to spatially guide the differentiation of NSCs [450, 456, 459]. Immobilization offers a number of advantages over soluble dosing. Immobilized molecules do not diffuse away from a scaffold like they would in soluble form and require smaller amounts of bioactive molecules over the course of treatment maintaining local concentrations, potentially maximizing cell interactions while reducing cost. Immobilization also allows for the creation of permanent gradients and this strategy has been utilized for small adhesion peptides [460–463] and more recently for growth factors [464, 465]. Consequently, it has become a preferred strategy employed to modify cell guidance and migration [460–463]. In achieving these effects, the presentation of proteins in gradients and/or 3D patterns can be highly beneficial since it more closely mimics how cells experience many proteins *in vivo*. Control of spatial distributions of proteins is key to achieving greater control of cell and tissue functions.

   Attachment peptides derived from ECM proteins have been immobilized to culture substrates for neural guidance to mimic ligand presentation during development and injury. A RGDC peptide

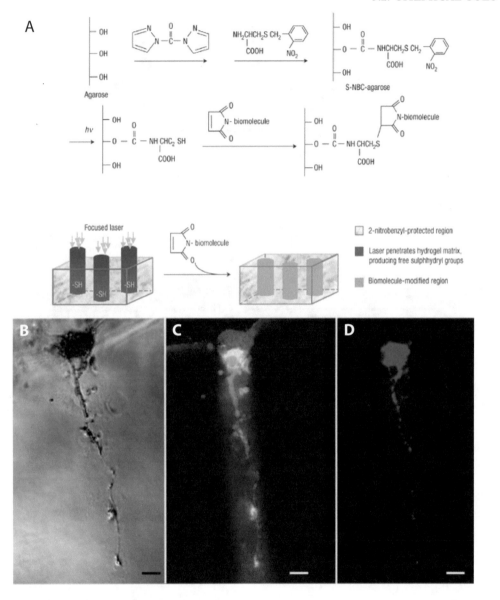

Figure 5.4: Chemically modified agarose hydrogels containing free sulfhydryl groups after exposure to a laser reacted with maleimido-terminated biomolecules (A). DRG cells extended into GRGDS-modified channels and not into the surrounding agarose matrix. Typical cell migration and axon extension is shown with: (B) optical microscopy; (C) confocal microscopy with cells stained red and the channel green; (D) fluorescent microscopy indicating cell migration into the GRGDS-modified channel via nuclear DAPI stain (blue). Images reprinted from [467].

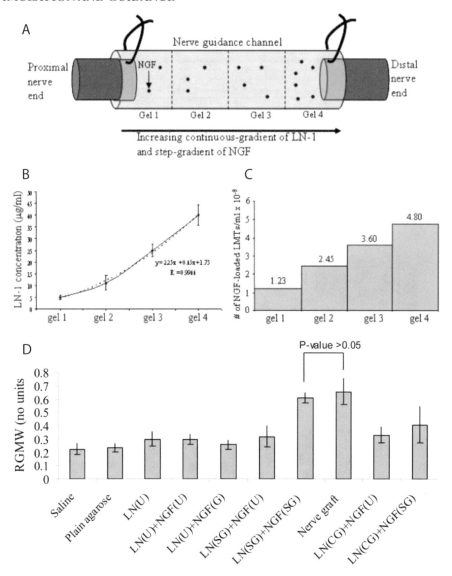

Figure 5.5: A nerve guidance channel scaffold for peripheral nerve repair made up of agarose with laminin (LN-1) and NGF (A). Within this scaffold the LN-1 gradient is continuous, as determined by ELISA (B), while NGF-loaded lipid microtubules (LMTs) are distributed in a step gradient (C). Implants with step gradients resulted in similar relative gastrocnemius muscle weight (RGMW) as nerve grafts and both treatments were significantly better than single and combinatorial treatments of LN-1 and NGF in uniform (U), step gradient (SG) and continuous gradient (CG) agarose scaffolds (D). Images reprinted from [100].

(found in laminin, collagen and fibronectin) and axonin-1, a cell adhesion protein, were patterned to a culture surface using photolithography techniques; this allowed *in vitro* neurite extension and network formation [466]. Similar studies by Luo *et al.* demonstrated that 2D photo-immobilized GRGDS enhanced local neurite density and elongation, as well as DRG extension, into immobilized GRGDS in 3D (Fig. 5.4) [467]. Photolithography paired with the development of multiphoton scanning microscopes [468–470], allows for even tighter control of immobilized factor concentration and spatial location, potentially providing axon guidance with submicron precision. Recent work from Yu *et al.* [464, 465] has utilized two-photon confocal patterning to tether NGF to chitosan surfaces. They demonstrated that immobilized NGF does encourage DRG axon extension *in vitro* and that axons can be guided by gradients of immobilized NGF.

Recombinant fusion proteins of growth factors and binding domains have been created previously to control immobilization to specific ECMs, biomaterials and cells. Fibronectin cell-binding domains have been incorporated in fusion proteins along with bFGF and EGF to stimulate vascularization and wound healing [471]. Collagen binding domains have also been incorporated into fusion proteins of bFGF [472], EGF [473, 474], PDGF [475], hepatocyte growth factor (HGF) [476] and NGF [108, 477] for targeted wound regeneration. Recently Sun *et al.* compared collagen binding NGF (CBD-NGF) to native NGF (NAT-NGF) in a crushed rat sciatic nerve model [108]. They found that CBD-NGF injection treatment enhanced remyelination of axons after crush injury. CBD-NGF resulted in better myelinated axons when compared to NAT-NGF and PBS sham treatments at 8 and 12 wks. Dodla and Ballamkonda [100] created nerve guidance scaffolds containing gradients of immobilized laminin and NGF in agarose hydrogels (Fig. 5.5) and showed that axons were able to bridge a 20 mm nerve gap after sciatic nerve injury using scaffolds with a continuous gradient of laminin and a step gradient of NGF in the same direction.

5.3   ELECTRICAL STIMULATION

Natural electrical activity of the nervous system has led to investigation on the effects of electrical stimulation, especially on neurons and subcellular behavior. The application of electric stimulants to neural cells is not a new idea by any means, and has been studied for over 30 years [478]. Despite its long history, mixed effects of applied electric fields still leave questions about their control over cell behavior. Neural stimulation *in vitro* could be used to promote desired cell behaviors such as alignment and outgrowth of processes. *In vivo*, electrical activity could be promoted by incorporating a conductive polymer mentioned in Chapter 3, such as PPy or poly(aniline).

Electric fields have been used to stimulate neurite outgrowth and have been found to align cells, giving them polarity. The presence of a direct current (DC) electric field causes growing axons *in vitro* to align, extend, and accelerate toward the cathode and to increase branching [479–483]. DRG in particular have been used extensively in studies of applied electric fields to study neurite sprouting, length, and alignment [484–486]. Neurites from DRG were observed to align in the direction of the field and extend significantly farther than control treatments, but these results were governed by surface properties [486]. Laminin coated surfaces induced the significant neurite lengthening, while collagen surfaces did not. In addition, substrate surfaces have shown variable results in neurite alignment and directionality in applied electric fields [296, 487].

Electrical stimulation has been investigated as a differentiation cue in mammalian neural stem cells. Ariza *et al.* recently found that DC electric fields of around 500 mV/mm applied across NSCs from the DG (dentate gyrus) induced significantly higher neuronal differentiation than alternating current (AC) electric fields or no electric field application [488]. In the experiments, DC voltage was applied *in vitro* for the first three days and the last day of culture but AC was applied the full 6 days. DC voltage did appear to cause a lower cell density and some cell death. Results suggest that DC electric fields might be selective for the neuronal phenotype and cause alignment of neurons from NSCs of the DG [488]. NSC migration in the presence of electric fields is of interest in inducing or guiding endogenous stem cells to injury sites. Adult rat NSCs did have directional migration toward the cathode over cells on which no electric field was applied [489].

Animal studies have shown that brief electric field stimulation (1 h/d) works as well or better than long stimulation (4 h/d to continuous) [490, 491]. Electric field stimulation has been shown to greatly enhance nerve regeneration in animal models, resulting in significant decreases in regeneration time. One hour a day of electrical stimulation has reduced healing time from 9 wk to only 3 wk in rat models [490].

Electric fields have shown considerable promise for enhanced neuronal development and regeneration; however, further work is needed to understand the optimal way to utilize electric fields for TE approaches, especially for methodologies involving stem cells.

CHAPTER 6

Concluding Remarks

Tissue engineering is poised to generate new techniques to replace or restore tissue damaged from injury or disease. The concept itself is not new, as was contemplated by several visionary scientists speculating on where basic biological knowledge gained decades ago would take future researchers [492]. Though significant progress has been made toward neural TE, especially in the PNS, there is still a long road ahead to formulate clinically relevant CNS solutions that achieve significant long-term functional benefits. Major issues in the area remain unresolved including global injury and disease models, functional assessment, complete eradication of unwanted immune response, an absolute cell source, and concrete mechanisms and techniques for physical, chemical, and electrical cues.

The continuing advancement of new technology and techniques coupled with the uncovering of basic knowledge brightens the CNS TE outlook and the creation of restorative therapies for devastating injuries such as SCI or TBI. Design of biomaterials from the ground up, including proper cellular and tissue functionalities, will allow for the creation of ideal brain and spinal cord regenerative constructs. In any TE strategy it is important to incorporate cues (chemical, physical, electrical, etc.) derived from native cellular microenvironment (Fig. 1.1) to instruct cells and tissues to predictably regenerate. The incorporation of multiple cues experienced during development and regeneration into CNS treatments has propelled neural regeneration and TE into new exciting frontiers. Neural TE researchers are progressing toward the hallmark milestones of enabling the restoration of speech after a TBI and empowering a paralyzed SCI patient to walk again.

Bibliography

[1] *Spinal cord injury facts and figures at a glance*. 2010, National Spinal Cord Injury Statistical Center: Birmingham, AL. p. 2. Cited on page(s) 1

[2] Faul, M., L. Xu, M.M. Wald, and V.G. Coronado, *Traumatic brain injury in the United States: emergency department visits, hospitalizations, and deaths*. 2010, Centers for Disease Control and Prevention, National Center for Injury Prevention and Control: Atlanta, GA. Cited on page(s) 1

[3] Jacobson, S. and E.M. Marcus, *Neuroanatomy for the neuroscientist*. 2008, Springer: Boston, MA. Cited on page(s) 3, 5, 7, 8, 9

[4] Levitan, I.B. and L.K. Kaczmarek, *The neuron cell and molecular biology*. Levitan, I. B. and L. K. Kaczmarek. The Neuron: Cell and Molecular Biology. Xii+450, p. Oxford University Press, Inc.: New York, New York, USA; Oxford, England, Uk. Illus, 1991: p. XII+450P. Cited on page(s) 3, 7, 8, 34

[5] Dityatev, A., C.I. Seidenbecher, and M. Schachner, *Compartmentalization from the outside: the extracellular matrix and functional microdomains in the brain*. Trends in Neurosciences, 2010. **33**(11): p. 503–512. DOI: 10.1016/j.tins.2010.08.003 Cited on page(s) 5, 8, 24

[6] Yu, L.M.Y., N.D. Leipzig, and M.S. Shoichet, *Promoting neuron adhesion and growth*. Materials Today, 2008. **11**(5): p. 36–43. DOI: 10.1016/S1369-7021(08)70088-9 Cited on page(s) 5, 53

[7] Milner, R. and I.L. Campbell, *The integrin family of cell adhesion molecules has multiple functions within the CNS*. Journal of Neuroscience Research, 2002. **69**(3): p. 286–291. DOI: 10.1002/jnr.10321 Cited on page(s) 5

[8] Hubert, T., et al., *Collagens in the developing and diseased nervous system*. Cellular and Molecular Life Sciences, 2009. **66**(7): p. 1223–1238. DOI: 10.1007/s00018-008-8561-9 Cited on page(s) 5

[9] Suzuki, T., et al., *Recent advances in the study of AMPA receptors*. Nihon yakurigaku zasshi. Folia pharmacologica Japonica, 2003. **122**(6): p. 515–26. DOI: 10.1254/fpj.122.515 Cited on page(s) 5

[10] Costa, C., et al., *Mapping of aggrecan, hyaluronic acid, heparan sulphate proteoglycans and aquaporin 4 in the central nervous system of the mouse.* Journal of Chemical Neuroanatomy, 2007. **33**(3): p. 111–123. DOI: 10.1016/j.jchemneu.2007.01.006 Cited on page(s) 5

[11] Ruoslahti, E., *Brain extracellular matrix.* Glycobiology, 1996. **6**(5): p. 489–492. DOI: 10.1093/glycob/6.5.489 Cited on page(s) 5

[12] Smith, D.H., *Stretch growth of integrated axon tracts: Extremes and exploitations.* Progress in Neurobiology, 2009. **89**(3): p. 231–239. DOI: 10.1016/j.pneurobio.2009.07.006 Cited on page(s) 7, 48

[13] Mallavarapu, A. and T. Mitchison, *Regulated actin cytoskeleton assembly at filopodium tips controls their extension and retraction.* Journal of Cell Biology, 1999. **146**(5): p. 1097–1106. DOI: 10.1083/jcb.146.5.1097 Cited on page(s) 7

[14] Huber, A.B., et al., *Signaling at the growth cone: Ligand-receptor complexes and the control of axon growth and guidance.* Annual Review of Neuroscience, 2003. **26**: p. 509–563. DOI: 10.1146/annurev.neuro.26.010302.081139 Cited on page(s) 7, 15, 55, 56, 57, 58

[15] Davis, L., et al., *Protein-synthesis within neuronal growth cones.* Journal of Neuroscience, 1992. **12**(12): p. 4867–4877. Cited on page(s) 7

[16] Luo, L. and D. O'Leary, *Axon retraction and degeneration in development and disease.* Annual Review of Neuroscience, 2005: p. 127–156. DOI: 10.1146/annurev.neuro.28.061604.135632 Cited on page(s) 8, 10, 11, 12, 14, 15

[17] Edgar, J.M. and K.A. Nave, *The role of CNS glia in preserving axon function.* Current Opinion in Neurobiology, 2009. **19**(5): p. 498–504. DOI: 10.1016/j.conb.2009.08.003 Cited on page(s) 8

[18] Ebneter, A., et al., *Microglial Activation in the Visual Pathway in Experimental Glaucoma: Spatiotemporal Characterization and Correlation with Axonal Injury.* Investigative Ophthalmology & Visual Science, 2010. **51**(12): p. 6448–6460. DOI: 10.1167/iovs.10-5284 Cited on page(s) 9

[19] Prewitt, C.M.F., et al., *Activated macrophage/microglial cells can promote the regeneration of sensory axons into the injured spinal cord.* Experimental Neurology, 1997. **148**(2): p. 433–443. DOI: 10.1006/exnr.1997.6694 Cited on page(s)

[20] Thored, P., et al., *Long-Term Accumulation of Microglia with Proneurogenic Phenotype Concomitant with Persistent Neurogenesis in Adult Subventricular Zone After Stroke.* Glia, 2009. **57**(8): p. 835–849. DOI: 10.1002/glia.20810 Cited on page(s) 9

[21] Cizkova, D., et al., *Response of Ependymal Progenitors to Spinal Cord Injury or Enhanced Physical Activity in Adult Rat.* Cellular and Molecular Neurobiology, 2009. **29**(6–7): p. 999-1013. DOI: 10.1007/s10571-009-9387-1 Cited on page(s) 9

[22] Hamilton, L.K., et al., *Cellular organization of the central canal ependymal zone, a niche of latent neural stem cells int the adult mammalian spinal cord.* Neuroscience, 2009. **164**(3): p. 1044–1056. DOI: 10.1016/j.neuroscience.2009.09.006 Cited on page(s) 42

[23] Hugnot, J.P. and R. Franzen, *The spinal cord ependymal region: A stem cell niche in the caudal central nervous system.* Frontiers in Bioscience-Landmark, 2011. **16**: p. 1044–1059. DOI: 10.2741/3734 Cited on page(s) 9, 42

[24] Basso, D.M., M.S. Beattie, and J.C. Bresnahan, *A sensitive and reliable locomotor rating-scale for open-field testing in rats.* Journal of Neurotrauma, 1995. **12**(1): p. 1–21. DOI: 10.1089/neu.1995.12.1 Cited on page(s) 9

[25] Lebedev, S.V., et al., *Exercise Tests and BBB Method for Evaluation of Motor Disorders in Rats after Contusion Spinal Injury.* Bulletin of Experimental Biology and Medicine, 2008. **146**(4): p. 489–494. DOI: 10.1007/s10517-009-0328-2 Cited on page(s) 9

[26] Colman, H., J. Nabekura, and J.W. Lichtman, *Alterations in synaptic strength preceding axon withdrawal.* Science, 1997. **275**(5298): p. 356–61. DOI: 10.1126/science.275.5298.356 Cited on page(s) 10

[27] Stent, G.S., *A physiological mechanism for Hebb's postulate of learning.* Proc Natl Acad Sci U S A, 1973. **70**(4): p. 997–1001. DOI: 10.1073/pnas.70.4.997 Cited on page(s) 10

[28] Solomon, F. and M. Magendantz, *Cytochalasin separates microtubule disassembly from loss of asymmetric morphology.* J Cell Biol, 1981. **89**(1): p. 157–61. DOI: 10.1083/jcb.89.1.157 Cited on page(s) 10

[29] Ahmad, F.J., et al., *Motor proteins regulate force interactions between microtubules and microfilaments in the axon.* Nat Cell Biol, 2000. **2**(5): p. 276–80. DOI: 10.1038/35010544 Cited on page(s) 10

[30] Nakayama, A.Y., M.B. Harms, and L. Luo, *Small GTPases Rac and Rho in the maintenance of dendritic spines and branches in hippocampal pyramidal neurons.* J Neurosci, 2000. **20**(14): p. 5329–38. Cited on page(s) 10

[31] Guan, K.L. and Y. Rao, *Signalling mechanisms mediating neuronal responses to guidance cues.* Nat Rev Neurosci, 2003. **4**(12): p. 941–56. DOI: 10.1038/nrn1254 Cited on page(s) 12

[32] He, Z. and V. Koprivica, *The Nogo signaling pathway for regeneration block.* Annu Rev Neurosci, 2004. **27**: p. 341–68. DOI: 10.1146/annurev.neuro.27.070203.144340 Cited on page(s) 12

[33] Coleman, M., *Axon degeneration mechanisms: Commonality amid diversity.* Nature Reviews Neuroscience, 2005: p. 889–898. DOI: 10.1038/nrn1788 Cited on page(s) 12, 14, 15

[34] Saxena, S. and P. Caroni, *Mechanisms of axon degeneration: From development to disease.* Progress in Neurobiology, 2007: p. 174–191. DOI: 10.1016/j.pneurobio.2007.07.007 Cited on page(s) 12

[35] Coleman, M.P. and M.R. Freeman, *Wallerian Degeneration, Wld(S), and Nmnat.* Annual Review of Neuroscience, Vol 33, 2010. **33**: p. 245–267.
DOI: 10.1146/annurev-neuro-060909-153248 Cited on page(s) 12, 14, 15

[36] Waller, A., *Experiments on the sections of glossopharyngeal and hypoglossal nerves of the frog and observations of the alterations produced thereby in the structures of their primitive fibers.* Philosophical Transactions of the Royal Society London, 1850. **140**. Cited on page(s) 12

[37] Benbassat, D. and M.E. Spira, *The survival of transected axonal segments of cultured aplysia neurons is prolonged by contact with intact nerve-cells.* European Journal of Neuroscience, 1994. **6**(10): p. 1605–1614. DOI: 10.1111/j.1460-9568.1994.tb00551.x Cited on page(s) 12

[38] Lunn, E., et al., *Absence of Wallerian degeneration does not hinder regeneration in peripheral-nerve.* European Journal of Neuroscience, 1989: p. 27–33.
DOI: 10.1111/j.1460-9568.1989.tb00771.x Cited on page(s) 12

[39] Beirowski, B., et al., *The progressive nature of Wallerian degeneration in wild-type and slow Wallerian degeneration (WldS) nerves.* BMC Neuroscience, 2005. **6**(February 1).
DOI: 10.1186/1471-2202-6-6 Cited on page(s) 12

[40] Raff, M.C., A.V. Whitmore, and J.T. Finn, *Neuroscience - Axonal self-destruction and neurodegeneration.* Science, 2002. **296**(5569): p. 868–871. DOI: 10.1126/science.1068613 Cited on page(s) 12'

[41] Glass, J.D., et al., *Prolonged survival of transected nerve-fibers in C57BL/OLA mice is an intrinsic characteristic of the axon.* Journal of Neurocytology, 1993. **22**(5): p. 311–321.
DOI: 10.1007/BF01195555 Cited on page(s) 12

[42] Wright, A.K., et al., *Synaptic Protection in the Brain of Wld(S) Mice Occurs Independently of Age but Is Sensitive to Gene-Dose.* Plos One, 2010. **5**(11): p. 9. DOI: 10.1371/journal.pone.0015108 Cited on page(s) 12

[43] Mack, T.G.A., et al., *Wallerian degeneration of injured axons and synapses is delayed by a Ube4b/Nmnat chimeric gene.* Nature Neuroscience, 2001. **4**(12): p. 1199–1206.
DOI: 10.1038/nn770 Cited on page(s) 12

[44] Bisby, M.A. and S. Chen, *Delayed Wallerian degeneration in sciatic-nerves of C57BL/OLA mice is associated with impaired regeneration of sensory axons*. Brain Research, 1990. **530**(1): p. 117–120. DOI: 10.1016/0006-8993(90)90666-Y Cited on page(s) 12

[45] Brown, M.C., E.R. Lunn, and V.H. Perry, *Consequences of slow Wallerian degeneration for regenerating motor and sensory axons*. Journal of Neurobiology, 1992. **23**(5): p. 521–536. DOI: 10.1002/neu.480230507 Cited on page(s) 12

[46] Martin, S.M., et al., *Wallerian degeneration of zebrafish trigeminal axons in the skin is required for regeneration and developmental pruning*. Development, 2010. **137**(23): p. 3985–3994. DOI: 10.1242/dev.053611 Cited on page(s) 12, 14

[47] Zhai, Q.W., et al., *Involvement of the ubiquitin-proteasome system in the early stages of Wallerian degeneration*. Neuron, 2003. **39**(2): p. 217–225. DOI: 10.1016/S0896-6273(03)00429-X Cited on page(s) 14

[48] Ghosh-Roy, A., et al., *Calcium and Cyclic AMP Promote Axonal Regeneration in Caenorhabditis elegans and Require DLK-1 Kinase*. Journal of Neuroscience, 2010. **30**(9): p. 3175–3183. DOI: 10.1523/JNEUROSCI.5464-09.2010 Cited on page(s) 14

[49] Ferri, A., et al., *Inhibiting axon degeneration and synapse loss attenuates apoptosis and disease progression in a mouse model of motoneuron disease*. Current Biology, 2003. **13**(8): p. 669–673. DOI: 10.1016/S0960-9822(03)00206-9 Cited on page(s) 15

[50] Horste, G.M.Z., et al., *The Wlds transgene reduces axon loss in a Charcot-Marie-Tooth disease 1A rat model and nicotinamide delays post-traumatic axonal degeneration*. Neurobiology of Disease, 2011. **42**(1): p. 1–8. DOI: 10.1016/j.nbd.2010.12.006 Cited on page(s) 15

[51] Beirowski, B., et al., *Mechanisms of Axonal Spheroid Formation in Central Nervous System Wallerian Degeneration*. Journal of Neuropathology and Experimental Neurology, 2010: p. 455–472. DOI: 10.1097/NEN.0b013e3181da84db Cited on page(s) 15, 37

[52] Mi, W.Q., et al., *The slow Wallerian degeneration gene, Wld(S), inhibits axonal spheroid pathology in gracile axonal dystrophy mice*. Brain, 2005. **128**: p. 405–416. DOI: 10.1093/brain/awh368 Cited on page(s) 15, 37

[53] Sajadi, A., B.L. Schneider, and P. Aebischer, *Wld(s)-mediated protection of dopaminergic fibers in an animal model of Parkinson disease*. Current Biology, 2004. **14**(4): p. 326–330. DOI: 10.1016/S0960-9822(04)00050-8 Cited on page(s) 15

[54] Samsam, M., et al., *The Wld(s) mutation delays robust loss of motor and sensory axons in a genetic model for myelin-related axonopathy*. Journal of Neuroscience, 2003. **23**(7): p. 2833–2839. Cited on page(s)

[55] Wang, M.S., et al., *The Wld(S) protein protects against axonal degeneration: A model of gene therapy for peripheral neuropathy.* Annals of Neurology, 2001. **50**(6): p. 773–779. DOI: 10.1002/ana.10039 Cited on page(s) 15

[56] Vargas, M.E. and B.A. Barres, *Why is Wallerian degeneration in the CNS so slow?* Annual Review of Neuroscience, 2007. **30**: p. 153–179. DOI: 10.1146/annurev.neuro.30.051606.094354 Cited on page(s) 15

[57] Yiu, G. and Z.G. He, *Glial inhibition of CNS axon regeneration.* Nature Reviews Neuroscience, 2006. **7**(8): p. 617–627. DOI: 10.1038/nrn1956 Cited on page(s) 14, 15, 16

[58] Huebner, E.A., et al., *A Multi-domain Fragment of Nogo-A Protein Is a Potent Inhibitor of Cortical Axon Regeneration via Nogo Receptor 1.* Journal of Biological Chemistry, 2011. **286**(20): p. 18026–18036. DOI: 10.1074/jbc.M110.208108 Cited on page(s) 15

[59] Wang, D., et al., *Neural stem cell transplantation with Nogo-66 receptor gene silencing to treat severe traumatic brain injury.* Neural Regeneration Research, 2011. **6**(10): p. 725–731. Cited on page(s) 15

[60] Gonzenbach, R.R. and M.E. Schwab, *Disinhibition of neurite growth to repair the injured adult CNS: Focusing on Nogo.* Cellular and Molecular Life Sciences, 2008. **65**(1): p. 161–176. DOI: 10.1007/s00018-007-7170-3 Cited on page(s) 15

[61] Li, M., et al., *Myelin-associated glycoprotein inhibits neurite/axon growth and causes growth cone collapse.* Journal of Neuroscience Research, 1996. **46**(4): p. 404–414. DOI: 10.1002/(SICI)1097-4547(19961115)46:4%3C404::AID-JNR2%3E3.0.CO;2-K Cited on page(s) 15

[62] Quarles, R.H., *A Hypothesis About the Relationship of Myelin-Associated Glycoprotein's Function in Myelinated Axons to its Capacity to Inhibit Neurite Outgrowth.* Neurochemical Research, 2009. **34**(1): p. 79–86. DOI: 10.1007/s11064-008-9668-y Cited on page(s) 15

[63] Ridet, J.L., et al., *Reactive astrocytes: cellular and molecular cues to biological function.* Trends in Neurosciences, 1997. **20**(12): p. 570–577. DOI: 10.1016/S0166-2236(97)01139-9 Cited on page(s) 16

[64] Menet, V., et al., *Axonal plasticity and functional recovery after spinal cord injury in mice deficient in both glial fibrillary acidic protein and vimentin genes.* Proceedings of the National Academy of Sciences of the United States of America, 2003. **100**(15): p. 8999–9004. DOI: 10.1073/pnas.1533187100 Cited on page(s) 16

[65] Niederost, B.P., et al., *Bovine CNS myelin contains neurite growth-inhibitory activity associated with chondroitin sulfate proteoglycans.* Journal of Neuroscience, 1999. **19**(20): p. 8979–8989. Cited on page(s) 16

[66] Bartanusz, V., et al., *The blood-spinal cord barrier: morphology and clinical implications.* Ann Neurol, 2011. **70**(2): p. 194–206. DOI: 10.1002/ana.22421 Cited on page(s) 16

[67] Sharma, H.S., *Pathophysiology of blood-spinal cord barrier in traumatic injury and repair.* Curr Pharm Des, 2005. **11**(11): p. 1353–89. DOI: 10.2174/1381612053507837 Cited on page(s)

[68] Pedersen, M.O., et al., *Cell death in the injured brain: roles of metallothioneins.* Prog Histochem Cytochem, 2009. **44**(1): p. 1–27. DOI: 10.1016/j.proghi.2008.10.002 Cited on page(s) 16

[69] Bonhomme, V., et al., *Neuron-specific enolase as a marker of invitro neuronal damage. 2. Investigation of the astrocyte protective effect against Kainate-induced neurotoxicity.* Journal of Neurosurgical Anesthesiology, 1993. **5**(2): p. 117–120. Cited on page(s) 16

[70] *Biomaterials science: An introduction to materials in medicine.* Biomaterials science: An introduction to materials in medicine, ed. B.D. Ratner, et al. 1996: Academic Press, Inc.; Academic Press Ltd. xi+484p. Cited on page(s) 17, 19, 21, 22, 25, 28

[71] Zhong, Y. and R.V. Bellamkonda, *Biomaterials for the central nervous system.* Journal of the Royal Society Interface, 2008. **5**(26): p. 957–975. DOI: 10.1098/rsif.2008.0071 Cited on page(s) 17, 21, 27

[72] Peyton, S.R., et al., *The emergence of ECM mechanics and cytoskeletal tension as important regulators of cell function.* Cell Biochemistry and Biophysics, 2007. **47**(2): p. 300–320. DOI: 10.1007/s12013-007-0004-y Cited on page(s) 17, 19

[73] Hejcl, A., et al., *Macroporous hydrogels based on 2-hydroxyethyl methacrylate. Part 6: 3D hydrogels with positive and negative surface charges and polyelectrolyte complexes in spinal cord injury repair.* Journal of Materials Science-Materials in Medicine, 2009. **20**(7): p. 1571–1577. DOI: 10.1007/s10856-009-3714-4 Cited on page(s) 17, 19, 28

[74] Nair, L.S. and C.T. Laurencin, *Biodegradable polymers as biomaterials.* Progress in Polymer Science, 2007. **32**(8–9): p. 762-798. DOI: 10.1016/j.progpolymsci.2007.05.017 Cited on page(s) 19, 24, 25

[75] Nisbet, D.R., et al., *Neural tissue engineering of the CNS using hydrogels.* Journal of Biomedical Materials Research Part B-Applied Biomaterials, 2008. **87B**(1): p. 251–263. DOI: 10.1002/jbm.b.31000 Cited on page(s) 21, 25, 27

[76] Straley, K.S., C.W.P. Foo, and S.C. Heilshorn, *Biomaterial Design Strategies for the Treatment of Spinal Cord Injuries.* Journal of Neurotrauma, 2010. **27**(1): p. 1–19. DOI: 10.1089/neu.2009.0948 Cited on page(s) 19, 24, 25, 27, 28

[77] Lee, K.Y. and D.J. Mooney, *Hydrogels for tissue engineering.* Chemical Reviews, 2001. **101**(7): p. 1869–1879. DOI: 10.1021/cr000108x Cited on page(s) 19

[78] West, J.L. and J.A. Hubbell, *Polymeric biomaterials with degradation sites for proteases involved in cell migration.* Macromolecules, 1999. **32**(1): p. 241–244. DOI: 10.1021/ma981296k Cited on page(s) 19, 21

[79] Lee, S.H., et al., *Poly(ethylene glycol) hydrogels conjugated with a collagenase-sensitive fluorogenic substrate to visualize collagenase activity during three-dimensional cell migration.* Biomaterials, 2007. **28**(20): p. 3163–3170. DOI: 10.1016/j.biomaterials.2007.03.004 Cited on page(s) 19, 21

[80] Kridel, S.J., et al., *Substrate hydrolysis by matrix metalloproteinase-9.* Journal of Biological Chemistry, 2001. **276**(23): p. 20572–20578. DOI: 10.1074/jbc.M100900200 Cited on page(s) 19

[81] Sottrup-Jensen, L. and H. Birkedal-Hansen, *Human fibroblast collagenase-alpha-macroglobulin interactions. Localization of cleavage sites in the bait regions of five mammalian alpha-macroglobulins.* J Biol Chem, 1989. **264**(1): p. 393–401. Cited on page(s)

[82] Turk, B.E., et al., *Determination of protease cleavage site motifs using mixture-based oriented peptide libraries.* Nature Biotechnology, 2001. **19**(7): p. 661–667. DOI: 10.1038/90273 Cited on page(s) 19

[83] Chau, Y., et al., *Incorporation of a matrix metalloproteinase-sensitive substrate into self-assembling peptides - A model for biofunctional scaffolds.* Biomaterials, 2008. **29**(11): p. 1713–1719. DOI: 10.1016/j.biomaterials.2007.11.046 Cited on page(s) 19

[84] Chau, Y., F.E. Tan, and R. Langer, *Synthesis and characterization of dextran-peptide-methotrexate conjugates for tumor targeting via mediation by matrix metalloproteinase II and matrix metalloproteinase IX.* Bioconjugate Chemistry, 2004. **15**(4): p. 931–941. DOI: 10.1021/bc0499174 Cited on page(s) 19, 21

[85] Tonti, G.A., et al., *Neural stem cells at the crossroads: MMPs may tell the way.* Int J Dev Biol, 2009. **53**(1): p. 1–17. DOI: 10.1387/ijdb.082573gt Cited on page(s) 21, 24

[86] Ogier, C., et al., *Matrix metalloproteinase-2 (MMP-2) regulates astrocyte motility in connection with the actin cytoskeleton and integrins.* Glia, 2006. **54**(4): p. 272–84. DOI: 10.1002/glia.20349 Cited on page(s) 21

[87] Zuo, J., et al., *Neuronal matrix metalloproteinase-2 degrades and inactivates a neurite-inhibiting chondroitin sulfate proteoglycan.* J Neurosci, 1998. **18**(14): p. 5203–11. Cited on page(s) 21

[88] Tayebjee, M.H., et al., *Effects of age, gender, ethnicity, diurnal variation and exercise on circulating levels of matrix metalloproteinases (MMP)-2 and-9, and their inhibitors, tissue inhibitors of matrix metalloproteinases (TIMP)-1 and-2.* Thrombosis Research, 2005. **115**(3): p. 205–210. DOI: 10.1016/j.thromres.2004.08.023 Cited on page(s) 21

[89] Katti, D.S., et al., *Toxicity, biodegradation and elimination of polyanhydrides.* Advanced Drug Delivery Reviews, 2002. **54**(7): p. 933–961. DOI: 10.1016/S0169-409X(02)00052-2 Cited on page(s) 21

[90] Leong, K.W., B.C. Brott, and R. Langer, *Bioerodible polyanhydrides as drug-carrier matrices. 1. Characterization, degradation, and release characteristics.* Journal of Biomedical Materials Research, 1985. **19**(8): p. 941–955. DOI: 10.1002/jbm.820190806 Cited on page(s) 21

[91] Baumann, M.D., et al., *Intrathecal delivery of a polymeric nanocomposite hydrogel after spinal cord injury.* Biomaterials, 2010. **31**(30): p. 7631–7639. DOI: 10.1016/j.biomaterials.2010.07.004 Cited on page(s) 21, 22, 27

[92] Pan, L., et al., *Viability and Differentiation of Neural Precursors on Hyaluronic Acid Hydrogel Scaffold.* Journal of Neuroscience Research, 2009. **87**(14): p. 3207–3220. DOI: 10.1002/jnr.22142 Cited on page(s) 21, 25

[93] Orive, G., et al., *Biomaterials for promoting brain protection, repair and regeneration.* Nature Reviews Neuroscience, 2009. **10**(9): p. 682-U47. DOI: 10.1038/nrn2685 Cited on page(s)

[94] Leipzig, N.D. and M.S. Shoichet, *The effect of substrate stiffness on adult neural stem cell behavior.* Biomaterials, 2009. **30**(36): p. 6867–6878. DOI: 10.1016/j.biomaterials.2009.09.002 Cited on page(s) 27, 48

[95] Saha, K., et al., *Substrate Modulus Directs Neural Stem Cell Behavior.* Biophysical Journal, 2008. **95**(9): p. 4426–4438. DOI: 10.1529/biophysj.108.132217 Cited on page(s) 21

[96] Cong Truc, H., N. Minh Khanh, and D.S. Lee, *Injectable Block Copolymer Hydrogels: Achievements and Future Challenges for Biomedical Applications.* Macromolecules, 2011. **44**(17): p. 6629–6636. DOI: 10.1021/ma201261m Cited on page(s) 22

[97] Jain, A., et al., *In situ gelling hydrogels for conformal repair of spinal cord defects, and local delivery of BDNF after spinal cord injury.* Biomaterials, 2006. **27**(3): p. 497–504. DOI: 10.1016/j.biomaterials.2005.07.008 Cited on page(s) 25

[98] Nguyen, M.K. and D.S. Lee, *Injectable biodegradable hydrogels.* Macromolecular bioscience, 2010. **10**(6): p. 563–79. DOI: 10.1002/mabi.200900402 Cited on page(s)

[99] Gupta, D., C.H. Tator, and M.S. Shoichet, *Fast-gelling injectable blend of hyaluronan and methylcellulose for intrathecal, localized delivery to the injured spinal cord.* Biomaterials, 2006. **27**(11): p. 2370–2379. DOI: 10.1016/j.biomaterials.2005.11.015 Cited on page(s) 22, 25, 27

[100] Dodla, M.C. and R.V. Bellamkonda, *Differences between the effect of anisotropic and isotropic laminin and nerve growth factor presenting scaffolds on nerve regeneration across long peripheral nerve gaps.* Biomaterials, 2008. **29**(1): p. 33–46. DOI: 10.1016/j.biomaterials.2007.08.045 Cited on page(s) 22, 62, 63

[101] Pfister, L.A., et al., *Nerve conduits and growth factor delivery in peripheral nerve repair.* Journal of the Peripheral Nervous System, 2007. **12**(2): p. 65–82. DOI: 10.1111/j.1529-8027.2007.00125.x Cited on page(s) 22

[102] Reneker, D.H. and I. Chun, *Nanometre diameter fibres of polymer, produced by electrospinning.* Nanotechnology, 1996. **7**(3): p. 216–223. DOI: 10.1088/0957-4484/7/3/009 Cited on page(s) 22

[103] Pham, Q.P., U. Sharma, and A.G. Mikos, *Electrospinning of polymeric nanofibers for tissue engineering applications: A review.* Tissue Engineering, 2006. **12**(5): p. 1197–1211. DOI: 10.1089/ten.2006.12.1197 Cited on page(s) 22, 51

[104] Xie, J.W., et al., *Electrospun nanofibers for neural tissue engineering.* Nanoscale, 2010. **2**(1): p. 35–44. DOI: 10.1039/b9nr00243j Cited on page(s) 22, 23, 51

[105] Gelain, F., et al., *Transplantation of Nanostructured Composite Scaffolds Results in the Regeneration of Chronically Injured Spinal Cords.* Acs Nano, 2011. **5**(1): p. 227–236. DOI: 10.1021/nn102461w Cited on page(s) 22, 28

[106] Lynn, A.K., I.V. Yannas, and W. Bonfield, *Antigenicity and immunogenicity of collagen.* Journal of Biomedical Materials Research Part B-Applied Biomaterials, 2004. **71B**(2): p. 343–354. DOI: 10.1002/jbm.b.30096 Cited on page(s) 24

[107] Fan, J., et al., *Linear Ordered Collagen Scaffolds Loaded with Collagen-Binding Neurotrophin-3 Promote Axonal Regeneration and Partial Functional Recovery after Complete Spinal Cord Transection.* Journal of Neurotrauma, 2010. **27**(9): p. 1671–1683. DOI: 10.1089/neu.2010.1281 Cited on page(s) 24

[108] Sun, W., et al., *The effect of collagen-binding NGF-beta on the promotion of sciatic nerve regeneration in a rat sciatic nerve crush injury model.* Biomaterials, 2009. **30**(27): p. 4649–56. DOI: 10.1016/j.biomaterials.2009.05.037 Cited on page(s) 24, 63

[109] Zhang, T., et al., *Three-dimensional gelatin and gelatin/hyaluronan hydrogel structures for traumatic brain injury.* Journal of Bioactive and Compatible Polymers, 2007. **22**(1): p. 19–29. DOI: 10.1177/0883911506074025 Cited on page(s) 24

[110] Zhang, H.Z., et al., *Gelatin-siloxane hybrid scaffolds with vascular endothelial growth factor induces brain tissue regeneration.* Current Neurovascular Research, 2008. **5**(2): p. 112–117. DOI: 10.2174/156720208784310204 Cited on page(s) 24

[111] Liu, T., et al., *Photochemical crosslinked electrospun collagen nanofibers: Synthesis, characterization and neural stem cell interactions.* Journal of Biomedical Materials Research Part A, 2010. **95A**(1): p. 276–282. DOI: 10.1002/jbm.a.32831 Cited on page(s) 25

[112] Timnak, A., et al., *Fabrication of nano-structured electrospun collagen scaffold intended for nerve tissue engineering.* Journal of Materials Science-Materials in Medicine, 2011. **22**(6): p. 1555–1567. DOI: 10.1007/s10856-011-4316-5 Cited on page(s) 25

[113] Wang, T.-W. and M. Spector, *Development of hyaluronic acid-based scaffolds for brain tissue engineering.* Acta Biomaterialia, 2009. **5**(7): p. 2371–2384. DOI: 10.1016/j.actbio.2009.03.033 Cited on page(s) 25

[114] Wei, Y.-T., et al., *Hyaluronic Acid Hydrogel Modified with Nogo-66 Receptor Antibody and Poly(L-Lysine) Enhancement of Adherence and Survival of Primary Hippocampal Neurons.* Journal of Bioactive and Compatible Polymers, 2009. **24**(3): p. 205–219. DOI: 10.1177/0883911509102266 Cited on page(s) 25

[115] Augst, A.D., H.J. Kong, and D.J. Mooney, *Alginate hydrogels as biomaterials.* Macromolecular Bioscience, 2006. **6**(8): p. 623–633. DOI: 10.1002/mabi.200600069 Cited on page(s) 25

[116] Gros, T., et al., *Regeneration of long-tract axons through sites of spinal cord injury using templated agarose scaffolds.* Biomaterials, 2010. **31**(26): p. 6719–6729. DOI: 10.1016/j.biomaterials.2010.04.035 Cited on page(s) 25, 26, 52

[117] Nisbet, D.R., et al., *Neurite infiltration and cellular response to electrospun polycaprolactone scaffolds implanted into the brain.* Biomaterials, 2009. **30**(27): p. 4573–4580. DOI: 10.1016/j.biomaterials.2009.05.011 Cited on page(s) 25

[118] Tripathi, A. and A. Kumar, *Multi-Featured Macroporous Agarose-Alginate Cryogel: Synthesis and Characterization for Bioengineering Applications.* Macromolecular Bioscience, 2011. **11**(1): p. 22–35. DOI: 10.1002/mabi.201000286 Cited on page(s) 25

[119] Bonino, C.A., et al., *Electrospinning alginate-based nanofibers: From blends to crosslinked low molecular weight alginate-only systems.* Carbohydrate Polymers, 2011. **85**(1): p. 111–119. DOI: 10.1016/j.carbpol.2011.02.002 Cited on page(s) 25

[120] Jeong, S.I., et al., *Electrospun Alginate Nanofibers with Controlled Cell Adhesion for Tissue Engineering.* Macromolecular Bioscience, 2010. **10**(8): p. 934–943. DOI: 10.1002/mabi.201000046 Cited on page(s) 25

[121] Crompton, K.E., et al., *Inflammatory response on injection of chitosan/GP to the brain.* Journal of Materials Science-Materials in Medicine, 2006. **17**(7): p. 633–639. DOI: 10.1007/s10856-006-9226-6 Cited on page(s) 27

[122] Sudarshan, N.R., D.G. Hoover, and D. Knorr, *Antibacterial action of chitosan.* Food Biotechnology, 1992. **6**(3): p. 257–272. DOI: 10.1080/08905439209549838 Cited on page(s) 27

[123] Crompton, K.E., et al., *Polylysine-functionalised thermoresponsive chitosan hydrogel for neural tissue engineering.* Biomaterials, 2007. **28**(3): p. 441–449. DOI: 10.1016/j.biomaterials.2006.08.044 Cited on page(s) 27

[124] Yu, L.M.Y., K. Kazazian, and M.S. Shoichet, *Peptide surface modification of methacrylamide chitosan for neural tissue engineering applications.* Journal of Biomedical Materials Research Part A, 2007. **82A**(1): p. 243–255. DOI: 10.1002/jbm.a.31069 Cited on page(s) 27

[125] Cao, Z., R.J. Gilbert, and W. He, *Simple Agarose-Chitosan Gel Composite System for Enhanced Neuronal Growth in Three Dimensions.* Biomacromolecules, 2009. **10**(10): p. 2954–2959. DOI: 10.1021/bm900670n Cited on page(s) 27

[126] Dainiak, M.B., et al., *Gelatin-fibrinogen cryogel dermal matrices for wound repair: Preparation, optimisation and in vitro study.* Biomaterials, 2010. **31**(1): p. 67–76. DOI: 10.1016/j.biomaterials.2009.09.029 Cited on page(s)

[127] Nikonorov, V.V., et al., *Synthesis and Characteristics of Cryogels of Chitosan Crosslinked by Glutaric Aldehyde.* Polymer Science Series A, 2010. **52**(8): p. 828–834. DOI: 10.1134/S0965545X10080092 Cited on page(s)

[128] Orrego, C.E. and J.S. Valencia, *Preparation and characterization of chitosan membranes by using a combined freeze gelation and mild crosslinking method.* Bioprocess and Biosystems Engineering, 2009. **32**(2): p. 197–206. DOI: 10.1007/s00449-008-0237-1 Cited on page(s) 27

[129] Jeong, S.I., et al., *Electrospun Chitosan–Alginate Nanofibers with In Situ Polyelectrolyte Complexation for Use as Tissue Engineering Scaffolds.* Tissue Engineering Part A, 2011. **17**(1–2): p. 59-70. DOI: 10.1089/ten.tea.2010.0086 Cited on page(s) 27

[130] Prabhakaran, M.P., et al., *Electrospun Biocomposite Nanofibrous Scaffolds for Neural Tissue Engineering.* Tissue Engineering Part A, 2008. **14**(11): p. 1787–1797. DOI: 10.1089/ten.tea.2007.0393 Cited on page(s) 27

[131] Cullen, D.K., et al., *In vitro neural injury model for optimization of tissue-engineered constructs.* Journal of Neuroscience Research, 2007. **85**(16): p. 3642–3651. DOI: 10.1002/jnr.21434 Cited on page(s) 27

[132] Tate, M.C., et al., *Biocompatibility of methylcellulose-based constructs designed for intracerebral gelation following experimental traumatic brain injury.* Biomaterials, 2001. **22**(10): p. 1113–1123. DOI: 10.1016/S0142-9612(00)00348-3 Cited on page(s) 27

[133] de Guzman, R.C., J.A. Loeb, and P.J. VandeVord, *Electrospinning of Matrigel to Deposit a Basal Lamina-Like Nanofiber Surface.* Journal of Biomaterials Science-Polymer Edition, 2010. **21**(8–9): p. 1081–1101. DOI: 10.1163/092050609X12457428936116 Cited on page(s) 27

[134] Bjugstad, K.B., et al., *Biocompatibility of poly(ethylene glycol)-based hydrogels in the brain: An analysis of the glial response across space and time.* Journal of Biomedical Materials Research Part A, 2010. **95A**(1): p. 79–91. DOI: 10.1002/jbm.a.32809 Cited on page(s) 27

[135] Piantino, J., et al., *An injectable, biodegradable hydrogel for trophic factor delivery enhances axonal rewiring and improves performance after spinal cord injury.* Experimental Neurology, 2006. **201**(2): p. 359–367. DOI: 10.1016/j.expneurol.2006.04.020 Cited on page(s) 28

[136] Yu, T.T. and M.S. Shoichet, *Guided cell adhesion and outgrowth in peptide-modified channels for neural tissue engineering.* Biomaterials, 2005. **26**(13): p. 1507–1514.
DOI: 10.1016/j.biomaterials.2004.05.012 Cited on page(s) 28

[137] Cigognini, D., et al., *Evaluation of Early and Late Effects into the Acute Spinal Cord Injury of an Injectable Functionalized Self-Assembling Scaffold.* Plos One, 2011. **6**(5): p. 15.
DOI: 10.1371/journal.pone.0019782 Cited on page(s) 28

[138] Gelain, F., A. Horii, and S.G. Zhang, *Designer self-assembling peptide scaffolds for 3-D tissue cell cultures and regenerative medicine.* Macromolecular Bioscience, 2007. **7**(5): p. 544–551.
DOI: 10.1002/mabi.200700033 Cited on page(s) 28

[139] Ghasemi-Mobarakeh, L., et al., *Application of conductive polymers, scaffolds and electrical stimulation for nerve tissue engineering.* Journal of Tissue Engineering and Regenerative Medicine, 2011. **5**(4): p. E17–E35. DOI: 10.1002/term.383 Cited on page(s) 28

[140] Lee, J.Y., J.-W. Lee, and C.E. Schmidt, *Neuroactive conducting scaffolds: nerve growth factor conjugation on active ester-functionalized polypyrrole.* Journal of the Royal Society Interface, 2009. **6**(38): p. 801–810. DOI: 10.1098/âĿ‹rsif.2008.0403 Cited on page(s) 29

[141] Shi, G., Z. Zhang, and M. Rouabhia, *The regulation of cell functions electrically using biodegradable polypyrrole-polylactide conductors.* Biomaterials, 2008. **29**(28): p. 3792–3798.
DOI: 10.1016/j.biomaterials.2008.06.010 Cited on page(s) 29

[142] Shi, G.X., et al., *A novel electrically conductive and biodegradable composite made of polypyrrole nanoparticles and polylactide.* Biomaterials, 2004. **25**(13): p. 2477–2488.
DOI: 10.1016/j.biomaterials.2003.09.032 Cited on page(s)

[143] Wang, Z.X., et al., *In vivo evaluation of a novel electrically conductive polypyrrole/poly(D,L-lactide) composite and polypyrrole-coated-poly(D,L-lactide-co-glycolide) membranes.* Journal of Biomedical Materials Research Part A, 2004. **70A**(1): p. 28–38. DOI: 10.1002/jbm.a.30047 Cited on page(s)

[144] Wang, Z.X., et al., *A biodegradable electrical bioconductor made of polypyrrole nanoparticle/poly(D,L-lactide) composite: A preliminary in vitro biostability study.* Journal of Biomedical Materials Research Part A, 2003. **66A**(4): p. 738–746.
DOI: 10.1002/jbm.a.10037 Cited on page(s) 29

[145] Baichwal, R.R., J.W. Bigbee, and G.H. Devries, *Macrophage-mediate myelin-related mitogenic factor for cultured Schwann-cells.* Proceedings of the National Academy of Sciences of the United States of America, 1988. **85**(5): p. 1701–1705. DOI: 10.1073/pnas.85.5.1701 Cited on page(s) 32

[146] Han, M.H., et al., *The role of Schwann-cells and macrophages in the removal of myelin during Wallerian degeneration.* Acta Histochemica Et Cytochemica, 1989. **22**(2): p. 161–172. DOI: 10.1267/ahc.22.161 Cited on page(s) 32

[147] Heumann, R., et al., *Differential regulation of messenger-RNA encoding nerve growth-factor and its receptor in rat sciatic-nerve during development, degeneration, and regeneration - role of macrophages.* Proceedings of the National Academy of Sciences of the United States of America, 1987. **84**(23): p. 8735–8739. DOI: 10.1073/pnas.84.23.8735 Cited on page(s) 33

[148] Meyer, M., et al., *Enhanced synthesis of brain-derived neurotrophic factor in the lesioned peripheral-nerve - different mechanisms are responsible for the regulation of BDNF and NGF messenger-RNA.* Journal of Cell Biology, 1992. **119**(1): p. 45–54. DOI: 10.1083/jcb.119.1.45 Cited on page(s) 33

[149] Gilmore, S.A. and D. Duncan, *On presence of peripheral-like nervous and connective tissue within irradiated spinal cord.* Anatomical Record, 1968. **160**(4): p. 675-&. DOI: 10.1002/ar.1091600403 Cited on page(s) 33

[150] Hirano, A., Zimmerma. Hm, and S. Levine, *Electron microscopic observations of peripheral myelin in a central nervous system lesion.* Acta Neuropathologica, 1969. **12**(4): p. 348-&. DOI: 10.1007/BF00809131 Cited on page(s)

[151] Raine, C.S., *Occurrence of Schwann-cells within normal central nervous-system.* Journal of Neurocytology, 1976. **5**(3): p. 371–380. DOI: 10.1007/BF01175122 Cited on page(s) 33

[152] Iwashita, Y. and W.F. Blakemore, *Areas of demyelination do not attract significant numbers of Schwann cells transplanted into normal white matter.* Glia, 2000. **31**(3): p. 232–240. DOI: 10.1002/1098-1136(200009)31:3%3C232::AID-GLIA40%3E3.0.CO;2-8 Cited on page(s) 33

[153] Iwashita, Y., et al., *Schwann cells transplanted into normal and x-irradiated adult white matter do not migrate extensively and show poor long-term survival.* Experimental Neurology, 2000. **164**(2): p. 292–302. DOI: 10.1006/exnr.2000.7440 Cited on page(s) 33

[154] Blakemore, W.F., *Limited remyelination of CNS axons by Schwann-cells transplanted into the sub-arachnoid space.* Journal of the Neurological Sciences, 1984. **64**(3): p. 265–276. DOI: 10.1016/0022-510X(84)90175-8 Cited on page(s) 33

[155] Blakemore, W.F., A.J. Crang, and R. Curtis, *The interaction of Schwann-cells with CNS axons in regions containing normal astrocytes.* Acta Neuropathologica, 1986. **71**(3–4): p. 295-300. DOI: 10.1007/BF00688052 Cited on page(s) 33

[156] Bachelin, C., et al., *Ectopic expression of polysialylated neural cell adhesion molecule in adult macaque Schwann cells promotes their migration and remyelination potential in the central nervous system.* Brain, 2010. **133**: p. 406–420. DOI: 10.1093/brain/awp256 Cited on page(s) 33

[157] Lavdas, A.A., et al., *Schwann cells genetically engineered to express PSA show enhanced migratory potential without impairment of their myelinating ability in vitro.* Glia, 2006. **53**(8): p. 868–878. DOI: 10.1002/glia.20340 Cited on page(s) 33

[158] Eccleston, P.A., K.R. Jessen, and R. Mirsky, *Control of peripheral glial-cell proliferation - a comparison of the division rates of enteric glia and Schwann-cells and their response to mitogens.* Developmental Biology, 1987. **124**(2): p. 409–417. DOI: 10.1016/0012-1606(87)90493-3 Cited on page(s) 33

[159] Jin, Y.-Q., et al., *Efficient Schwann cell purification by differential cell detachment using multiplex collagenase treatment.* Journal of Neuroscience Methods, 2008. **170**(1): p. 140–148. DOI: 10.1016/j.jneumeth.2008.01.003 Cited on page(s) 33

[160] Singh, A.K., et al., *Schwann cell culture from the adult animal sciatic nerve: Technique and review.* Journal of Clinical Neuroscience, 1996. **3**(1): p. 69–74. DOI: 10.1016/S0967-5868(96)90086-7 Cited on page(s) 33

[161] Funk, D., C. Fricke, and B. Schlosshauer, *Aging Schwann cells in vitro.* European Journal of Cell Biology, 2007. **86**(4): p. 207–219. DOI: 10.1016/j.ejcb.2006.12.006 Cited on page(s) 33

[162] Langford, L.A., S. Porter, and R.P. Bunge, *Immortalized rat Schwann-cells produce tumors invivo.* Journal of Neurocytology, 1988. **17**(4): p. 521–529. DOI: 10.1007/BF01189807 Cited on page(s) 33

[163] Lakatos, A., R.J.M. Franklin, and S.C. Barnett, *Olfactory ensheathing cells and Schwann cells differ in their in vitro interactions with astrocytes.* Glia, 2000. **32**(3): p. 214–225. DOI: 10.1002/1098-1136(200012)32:3%3C214::AID-GLIA20%3E3.0.CO;2-7 Cited on page(s) 33

[164] Imaizumi, T., et al., *Xenotransplantation of transgenic pig olfactory ensheathing cells promotes axonal regeneration in rat spinal cord.* Nature Biotechnology, 2000. **18**(9): p. 949–953. DOI: 10.1038/79432 Cited on page(s) 33

[165] Lu, P., et al., *Olfactory ensheathing cells do not exhibit unique migratory or axonal growth-promoting properties after spinal cord injury.* Journal of Neuroscience, 2006. **26**(43): p. 11120–11130. DOI: 10.1523/JNEUROSCI.3264-06.2006 Cited on page(s) 33, 34

[166] Keyvan-Fouladi, N., G. Raisman, and Y. Li, *Functional repair of the corticospinal tract by delayed transplantation of olfactory ensheathing cells in adult rats.* Journal of Neuroscience, 2003. **23**(28): p. 9428–9434. Cited on page(s) 33

[167] Ramon-Cueto, A., et al., *Functional recovery of paraplegic rats and motor axon regeneration in their spinal cords by olfactory ensheathing glia.* Neuron, 2000. **25**(2): p. 425–435. DOI: 10.1016/S0896-6273(00)80905-8 Cited on page(s) 33

[168] Cao, L., et al., *Olfactory ensheathing cells genetically modified to secrete GDNF to promote spinal cord repair.* Brain, 2004. **127**: p. 535–549. DOI: 10.1093/brain/awh072 Cited on page(s) 34

[169] Ruitenberg, M.J., et al., *Ex vivo adenoviral vector-mediated neurotrophin gene transfer to olfactory ensheathing glia: Effects on rubrospinal tract regeneration, lesion size, and functional recovery after implantation in the injured rat spinal cord.* Journal of Neuroscience, 2003. **23**(18): p. 7045–7058. Cited on page(s) 34

[170] Carlstedt, T., C.J. Dalsgaard, and C. Molander, *Regrowth of lesioned dorsal-root nerve-fibers into the spinal-cord of neonatal rats.* Neuroscience Letters, 1987. **74**(1): p. 14–18. DOI: 10.1016/0304-3940(87)90043-7 Cited on page(s) 34

[171] Lankford, K.L., et al., *Olfactory Ensheathing Cells Exhibit Unique Migratory, Phagocytic, and Myelinating Properties in the X-Irradiated Spinal Cord not Shared by Schwann Cells.* Glia, 2008. **56**(15): p. 1664–1678. DOI: 10.1002/glia.20718 Cited on page(s)

[172] Sasaki, M., et al., *Identified olfactory ensheathing cells transplanted into the transected dorsal funiculus bridge the lesion and form myelin.* Journal of Neuroscience, 2004. **24**(39): p. 8485–8493. DOI: 10.1523/JNEUROSCI.1998-04.2004 Cited on page(s) 34

[173] Au, E., et al., *SPARC from olfactory ensheathing cells stimulates Schwann cells to promote neurite outgrowth and enhances spinal cord repair.* Journal of Neuroscience, 2007. **27**(27): p. 7208–7221. DOI: 10.1523/JNEUROSCI.0509-07.2007 Cited on page(s) 34

[174] Fouad, K., et al., *Combining Schwann cell bridges and olfactory-ensheathing glia grafts with chondroitinase promotes locomotor recovery after complete transection of the spinal cord.* Journal of Neuroscience, 2005. **25**(5): p. 1169–1178. DOI: 10.1523/JNEUROSCI.3562-04.2005 Cited on page(s) 34

[175] Joannides, A.J. and S. Chandran, *Human embryonic stem cells: An experimental and therapeutic resource for neurological disease.* Journal of the Neurological Sciences, 2008. **265**(1–2): p. 84–88. DOI: 10.1016/j.jns.2007.09.008 Cited on page(s) 34

[176] Rogers, C.D., S.A. Moody, and E.S. Casey, *Neural Induction and Factors That Stabilize a Neural Fate.* Birth Defects Research Part C-Embryo Today-Reviews, 2009. **87**(3): p. 249–262. DOI: 10.1002/bdrc.20157 Cited on page(s) 34

[177] Maden, M., *Retinoic acid in the development, regeneration and maintenance of the nervous system.* Nature Reviews Neuroscience, 2007. **8**(10): p. 755–765. DOI: 10.1038/nrn2212 Cited on page(s) 34

[178] Villanueva, S., et al., *Posteriorization by FGF, Wnt, and retinoic acid is required for neural crest induction.* Developmental Biology, 2002. **241**(2): p. 289–301. DOI: 10.1006/dbio.2001.0485 Cited on page(s) 34

[179] Jessell, T.M., *Neuronal specification in the spinal cord: Inductive signals and transcriptional codes.* Nature Reviews Genetics, 2000. **1**(1): p. 20–29. DOI: 10.1038/35049541 Cited on page(s) 34

[180] Liem, K.F., T.M. Jessell, and J. Briscoe, *Regulation of the neural patterning activity of sonic hedgehog by secreted BMP inhibitors expressed by notochord and somites.* Development, 2000. **127**(22): p. 4855–4866. Cited on page(s) 34

[181] Delfino-Machin, M., et al., *The proliferating field of neural crest stem cells.* Developmental Dynamics, 2007. **236**(12): p. 3242–3254. DOI: 10.1002/dvdy.21314 Cited on page(s) 34, 44

[182] Le Douarin, N.M., G.W. Calloni, and E. Dupin, *The stem cells of the neural crest.* Cell Cycle, 2008. **7**(8): p. 1013–1019. DOI: 10.4161/cc.7.8.5641 Cited on page(s)

[183] Nagoshi, N., et al., *Neural Crest-Derived Stem Cells Display a Wide Variety of Characteristics.* Journal of Cellular Biochemistry, 2009. **107**(6): p. 1046–1052. DOI: 10.1002/jcb.22213 Cited on page(s) 34, 44

[184] Li, X.J., et al., *Specification of motoneurons from human embryonic stem cells.* Nature Biotechnology, 2005. **23**(2): p. 215–221. DOI: 10.1038/nbt1063 Cited on page(s) 35

[185] Li, X.J., et al., *Directed differentiation of ventral spinal progenitors and motor neurons from human embryonic stem cells by small molecules.* Stem Cells, 2008. **26**(4): p. 886–893. DOI: 10.1634/stemcells.2007-0620 Cited on page(s)

[186] Wichterle, H., et al., *Directed differentiation of embryonic stem cells into motor neurons.* Cell, 2002. **110**(3): p. 385–397. DOI: 10.1016/S0092-8674(02)00835-8 Cited on page(s) 35

[187] Bain, G., et al., *Retinoic acid promotes neural and represses mesodermal gene expression in mouse embryonic stem cells in culture.* Biochemical and Biophysical Research Communications, 1996. **223**(3): p. 691–694. DOI: 10.1006/bbrc.1996.0957 Cited on page(s) 35

[188] Bibel, M., et al., *Differentiation of mouse embryonic stem cells into a defined neuronal lineage.* Nature Neuroscience, 2004. **7**(9): p. 1003–1009. DOI: 10.1038/nn1301 Cited on page(s) 35, 36

[189] Freude, K.K., et al., *Soluble Amyloid Precursor Protein Induces Rapid Neural Differentiation of Human Embryonic Stem Cells*. Journal of Biological Chemistry, 2011. **286**(27): p. 24264–24274. DOI: 10.1074/jbc.M111.227421 Cited on page(s) 37

[190] Boido, M., et al., *Embryonic and adult stem cells promote raphespinal axon outgrowth and improve functional outcome following spinal hemisection in mice*. European Journal of Neuroscience, 2009. **30**(5): p. 833–846. DOI: 10.1111/j.1460-9568.2009.06879.x Cited on page(s) 37

[191] Perrin, F.E., et al., *Grafted Human Embryonic Progenitors Expressing Neurogenin-2 Stimulate Axonal Sprouting and Improve Motor Recovery after Severe Spinal Cord Injury*. Plos One, 2010. **5**(12): p. 7. DOI: 10.1371/journal.pone.0015914 Cited on page(s) 37

[192] Takahashi, K. and S. Yamanaka, *Induction of pluripotent stem cells from mouse embryonic and adult fibroblast cultures by defined factors*. Cell, 2006. **126**(4): p. 663–676. DOI: 10.1016/j.cell.2006.07.024 Cited on page(s) 37

[193] Kim, J.B., et al., *Pluripotent stem cells induced from adult neural stem cells by reprogramming with two factors*. Nature, 2008. **454**(7204): p. 646–U54. DOI: 10.1038/nature07061 Cited on page(s) 37

[194] Schatzlein, A.G., *Non-viral vectors in cancer gene therapy: principles and progress*. Anti-Cancer Drugs, 2001. **12**(4): p. 275–304. DOI: 10.1097/00001813-200104000-00001 Cited on page(s) 37, 38

[195] Kaji, K., et al., *Virus-free induction of pluripotency and subsequent excision of reprogramming factors*. Nature, 2009. **458**(7239): p. 771–U112. DOI: 10.1038/nature07864 Cited on page(s) 38

[196] Lee, C.H., et al., *The generation of iPS cells using non-viral magnetic nanoparticlebased transfection*. Biomaterials, 2011. **32**(28): p. 6683–91. DOI: 10.1016/j.biomaterials.2011.05.070 Cited on page(s)

[197] Montserrat, N., et al., *Simple Generation of Human Induced Pluripotent Stem Cells Using Poly-beta-amino Esters As the Non-viral Gene Delivery System*. Journal of Biological Chemistry, 2011. **286**(14). DOI: 10.1074/jbc.M110.168013 Cited on page(s) 38

[198] Okita, K., et al., *Generation of Mouse Induced Pluripotent Stem Cells Without Viral Vectors*. Science, 2008. **322**(5903): p. 949–953. DOI: 10.1126/science.1164270 Cited on page(s) 39

[199] Khalil, I.A., et al., *Uptake pathways and subsequent intracellular trafficking in nonviral gene delivery*. Pharmacological Reviews, 2006. **58**(1): p. 32–45. DOI: 10.1124/pr.58.1.8 Cited on page(s) 39

[200] Ma, H. and S.L. Diamond, *Nonviral gene therapy and its delivery systems.* Current Pharmaceutical Biotechnology, 2001. **2**(1): p. 1–17. DOI: 10.2174/1389201013378770 Cited on page(s) 39

[201] Malgrange, B., et al., *Using human pluripotent stem cells to untangle neurodegenerative disease mechanisms.* Cellular and Molecular Life Sciences, 2011. **68**(4): p. 635–649. DOI: 10.1007/s00018-010-0557-6 Cited on page(s) 38, 39

[202] Dimos, J.T., et al., *Induced pluripotent stem cells generated from patients with ALS can be differentiated into motor neurons.* Science, 2008. **321**(5893): p. 1218–1221. DOI: 10.1126/science.1158799 Cited on page(s) 39

[203] Park, I.H., et al., *Disease-specific induced pluripotent stem cells.* Cell, 2008. **134**(5): p. 877–886. DOI: 10.1016/j.cell.2008.07.041 Cited on page(s)

[204] Soldner, F., et al., *Parkinson's Disease Patient-Derived Induced Pluripotent Stem Cells Free of Viral Reprogramming Factors.* Cell, 2009. **136**(5): p. 964–977. DOI: 10.1016/j.cell.2009.02.013 Cited on page(s) 39

[205] Weiss, S., et al., *Multipotent CNS stem cells are present in the adult mammalian spinal cord and ventricular neuroaxis.* Journal of Neuroscience, 1996. **16**(23): p. 7599–7609. DOI: 10.1016/S0959-4388(99)80017-8 Cited on page(s) 39, 42

[206] Lois, C. and A. Alvarezbuylla, *Proliferating subventricular zone cells in the adult mammalian forebrain can differentiate into neurons and glia.* Proceedings of the National Academy of Sciences of the United States of America, 1993. **90**(5): p. 2074–2077. DOI: 10.1073/pnas.90.5.2074 Cited on page(s) 39

[207] Gage, F.H., et al., *Multipotent progenitor cells in the adult dentate gyrus.* Journal of Neurobiology, 1998. **36**(2): p. 249–266. DOI: 10.1002/(SICI)1097-4695(199808)36:2%3C249::AID-NEU11%3E3.0.CO;2-9 Cited on page(s) 39

[208] Altman, J. and G.D. Das, *Autoradiographic and histological evidence of postinatal hippocampal neurogenesis in rats.* Journal of Comparative Neurology, 1965. **124**(3): p. 319-&. DOI: 10.1002/cne.901240303 Cited on page(s) 39

[209] Eriksson, P.S., et al., *Neurogenesis in the adult human hippocampus.* Nature Medicine, 1998. **4**(11): p. 1313–1317. DOI: 10.1038/3305 Cited on page(s)

[210] Gould, E., et al., *Hippocampal neurogenesis in adult Old World primates.* Proceedings of the National Academy of Sciences of the United States of America, 1999. **96**(9): p. 5263–5267. DOI: 10.1073/pnas.96.9.5263 Cited on page(s)

[211] Kornack, D.R. and P. Rakic, *Continuation of neurogenesis in the hippocampus of the adult macaque monkey.* Proceedings of the National Academy of Sciences of the United States of America, 1999. **96**(10): p. 5768–5773. DOI: 10.1073/pnas.96.10.5768 Cited on page(s) 39

[212] Jankovski, A. and C. Sotelo, *Subventricular zone olfactory bulb migratory pathway in the adult mouse: Cellular composition and specificity as determined by heterochronic and heterotopic transplantation.* Journal of Comparative Neurology, 1996. **371**(3): p. 376–396. DOI: 10.1002/(SICI)1096-9861(19960729)371:3%3C376::AID-CNE3%3E3.3.CO;2-V Cited on page(s) 39

[213] Nissant, A. and M. Pallotto, *Integration and maturation of newborn neurons in the adult olfactory bulb - from synapses to function.* European Journal of Neuroscience, 2011. **33**(6): p. 1069–1077. DOI: 10.1111/j.1460-9568.2011.07605.x Cited on page(s)

[214] Pignatelli, A., C. Gambardella, and O. Belluzzi, *Neurogenesis in the adult olfactory bulb.* Neural Regeneration Research, 2011. **6**(8): p. 575–600. Cited on page(s)

[215] Morshead, C.M. and D. Vanderkooy, *Postmitotic death is the fate of constitutively proliferating cells in the subependymal layer of the adult-mouse brain.* Journal of Neuroscience, 1992. **12**(1): p. 249–256. Cited on page(s) 39

[216] Suhonen, J.O., et al., *Differentiation of adult hippocampus-derived progenitors into olfactory neurons in vivo.* Nature, 1996. **383**(6601): p. 624–627. DOI: 10.1038/383624a0 Cited on page(s) 39

[217] Nomura, H., et al., *Extramedullary chitosan channels promote survival of transplanted neural stem and progenitor cells and create a tissue bridge after complete spinal cord transection.* Tissue Engineering Part A, 2008. **14**(5): p. 649–665. DOI: 10.1089/tea.2007.0180 Cited on page(s) 39, 41, 42

[218] Rice, A.C., et al., *Proliferation and neuronal differentiation of mitotically active cells following traumatic brain injury.* Experimental Neurology, 2003. **183**(2): p. 406–417. DOI: 10.1016/S0014-4886(03)00241-3 Cited on page(s) 39

[219] Bye, N., et al., *Neurogenesis and Glial Proliferation Are Stimulated Following Diffuse Traumatic Brain Injury in Adult Rats.* Journal of Neuroscience Research, 2011. **89**(7): p. 986–1000. DOI: 10.1002/jnr.22635 Cited on page(s) 42

[220] Nait-Oumesmar, B., et al., *Progenitor cells of the adult mouse subventricular zone proliferate, migrate and differentiate into oligodendrocytes after demyelination.* European Journal of Neuroscience, 1999. **11**(12): p. 4357–4366. DOI: 10.1046/j.1460-9568.1999.00873.x Cited on page(s) 42

[221] Kulbatski, I., et al., *Endogenous and exogenous CNS derived stem/progenitor cell approaches for neurotrauma.* Current Drug Targets, 2005. **6**(1): p. 111–126. DOI: 10.2174/1389450053345037 Cited on page(s) 42

[222] Aberg, M.A.I., et al., *Peripheral infusion of IGF-I selectively induces neurogenesis in the adult rat hippocampus.* Journal of Neuroscience, 2000. **20**(8): p. 2896–2903. Cited on page(s) 42

[223] Benraiss, A., et al., *Adenoviral brain-derived neurotrophic factor induces both neostriatal and olfactory neuronal recruitment from endogenous progenitor cells in the adult forebrain.* Journal of Neuroscience, 2001. **21**(17): p. 6718–6731. Cited on page(s) 42

[224] Bai, Y., et al., *Ectopic expression of angiopoietin-1 promotes neuronal differentiation in neural progenitor cells through the Akt pathway.* Biochemical and Biophysical Research Communications, 2009. **378**(2): p. 296–301. DOI: 10.1016/j.bbrc.2008.11.052 Cited on page(s) 42

[225] Bath, K.G. and F.S. Lee, *Neurotrophic Factor Control of Adult SVZ Neurogenesis.* Developmental Neurobiology, 2010. **70**(5): p. 339–349. DOI: 10.1002/dneu.20781 Cited on page(s)

[226] Rosa, A.I., et al., *The Angiogenic Factor Angiopoietin-1 Is a Proneurogenic Peptide on Subventricular Zone Stem/Progenitor Cells.* Journal of Neuroscience, 2010. **30**(13): p. 4573–4584. DOI: 10.1523/JNEUROSCI.5597-09.2010 Cited on page(s)

[227] Wong, G., Y. Goldshmit, and A.M. Turnley, *Interferon-gamma but not TNF alpha promotes neuronal differentiation and neurite outgrowth of murine adult neural stem cells.* Experimental Neurology, 2004. **187**(1): p. 171–177. DOI: 10.1016/j.expneurol.2004.01.009 Cited on page(s) 42

[228] Zahir, T., et al., *Neural Stem/Progenitor Cells Differentiate In Vitro to Neurons by the Combined Action of Dibutyryl cAMP and Interferon-gamma.* Stem Cells and Development, 2009. **18**(10): p. 1423–1432. DOI: 10.1089/scd.2008.0412 Cited on page(s) 42

[229] Arsenijevic, Y. and S. Weiss, *Insulin-like growth factor-I is a differentiation factor for postmitotic CNS stem cell-derived neuronal precursors: Distinct actions from those of brain-derived neurotrophic factor.* Journal of Neuroscience, 1998. **18**(6): p. 2118–2128. Cited on page(s) 42

[230] Arsenijevic, Y., et al., *Insulin-like growth factor-1 is necessary for neural stem cell proliferation and demonstrates distinct actions of epidermal growth factor and fibroblast growth factor-2.* Journal of Neuroscience, 2001. **21**(18): p. 7194–7202. Cited on page(s) 42

[231] Kim, S.J., et al., *Interferon-gamma promotes differentiation of neural progenitor cells via the JNK pathway.* Neurochemical Research, 2007. **32**(8): p. 1399–1406. DOI: 10.1007/s11064-007-9323-z Cited on page(s) 42

[232] Adrian, E.K., Jr. and B.E. Walker, *Incorporation of thymidine-H3 by cells in normal and injured mouse spinal cord.* Journal of neuropathology and experimental neurology, 1962. **21**: p. 597–609. DOI: 10.1097/00005072-196210000-00007 Cited on page(s) 42

[233] Matthews, M.A., M.F. Stonge, and C.L. Faciane, *Electron-microscopic analysis of abnormal ependymal cell-proliferation and envelopment of sprouting axons following spinal-cord transection in the rat.* Acta Neuropathologica, 1979. **45**(1): p. 27–36. DOI: 10.1007/BF00691801 Cited on page(s) 42

[234] Meletis, K., et al., *Spinal cord injury reveals multilineage differentiation of ependymal cells.* Plos Biology, 2008. **6**(7): p. 1494–1507. DOI: 10.1371/journal.pbio.0060182 Cited on page(s)

[235] Rakic, P. and R.L. Sidman, *Subcommissural organ and adjacent ependyma - autoradiographic study of their origin in mouse brain.* American Journal of Anatomy, 1968. **122**(2): p. 317-&. DOI: 10.1002/aja.1001220210 Cited on page(s) 42

[236] Attar, A., et al., *Electron microscopic study of the progeny of ependymal stem cells in the normal and injured spinal cord.* Surgical Neurology, 2005. **64**: p. 28–32. DOI: 10.1016/j.surneu.2005.07.057 Cited on page(s) 42

[237] Kojima, A. and C.H. Tator, *Intrathecal administration of epidermal growth factor and fibroblast growth factor 2 promotes ependymal proliferation and functional recovery after spinal cord injury in adult rats.* Journal of Neurotrauma, 2002. **19**(2): p. 223–238. DOI: 10.1089/08977150252806974 Cited on page(s) 42

[238] Ayuso-Sacido, A., et al., *Long-term expansion of adult human brain subventricular zone precursors.* Neurosurgery, 2008. **62**(1): p. 223–229. DOI: 10.1227/01.NEU.0000311081.50648.4C Cited on page(s) 42

[239] Kuhn, H.G., H. DickinsonAnson, and F.H. Gage, *Neurogenesis in the dentate gyrus of the adult rat: Age-related decrease of neuronal progenitor proliferation.* Journal of Neuroscience, 1996. **16**(6): p. 2027–2033. Cited on page(s) 42

[240] Barry, F.P. and J.M. Murphy, *Mesenchymal stem cells: clinical applications and biological characterization.* International Journal of Biochemistry & Cell Biology, 2004. **36**(4): p. 568–584. DOI: 10.1016/j.biocel.2003.11.001 Cited on page(s) 43

[241] Alexanian, A.R., *An efficient method for generation of neural-like cells from adult human bone marrow-derived mesenchymal stem cells.* Regenerative Medicine, 2010. **5**(6): p. 891–900. DOI: 10.2217/rme.10.67 Cited on page(s) 43

[242] Galmiche, M.C., et al., *Stromal cells from human long-term marrow cultures are mesenchymal cells that differentiate following a vascular smooth-muscle differentiation pathway.* Blood, 1993. **82**(1): p. 66–76. Cited on page(s)

[243] Jackson, K.A., et al., *Regeneration of ischemic cardiac muscle and vascular endothelium by adult stem cells.* Journal of Clinical Investigation, 2001. **107**(11): p. 1395–1402. DOI: 10.1172/JCI12150 Cited on page(s)

[244] Theise, N.D., et al., *Derivation of hepatocytes from bone marrow cells in mice after radiation-induced myeloablation.* Hepatology, 2000. **31**(1): p. 235–240. DOI: 10.1002/hep.510310135 Cited on page(s)

[245] Zimmet, J.M. and J.M. Hare, *Emerging role for bone marrow derived mesenchymal stem cells in myocardial regenerative therapy.* Basic Research in Cardiology, 2005. **100**(6): p. 471–481. DOI: 10.1007/s00395-005-0553-4 Cited on page(s)

[246] Kashani, I.R., et al., *Schwann-like cell differentiation from rat bone marrow stem cells.* Archives of Medical Science, 2011. **7**(1): p. 45–52. DOI: 10.5114/aoms.2011.20603 Cited on page(s) 43

[247] Battula, V.L., et al., *Human placenta and bone marrow derived MSC cultured in serum-free, b-FGF-containing medium express cell surface frizzled-9 and SSEA-4 and give rise to multilinelage differentiation.* Differentiation, 2007. **75**(4): p. 279–291. DOI: 10.1111/j.1432-0436.2006.00139.x Cited on page(s) 43

[248] Datta, I., et al., *Neuronal plasticity of human Wharton's jelly mesenchymal stromal cells to the dopaminergic cell type compared with human bone marrow mesenchymal stromal cells.* Cytotherapy, 2011. **13**(8): p. 918–32. DOI: 10.3109/14653249.2011.579957 Cited on page(s)

[249] Lu, L.-L., et al., *Isolation and characterization of human umbilical cord mesenchymal stem cells with hematopoiesis-supportive function and other potentials.* Haematologica-the Hematology Journal, 2006. **91**(8): p. 1017–1026. Cited on page(s) 43

[250] Walker, P.A., et al., *Progenitor cell therapy for the treatment of central nervous system injury: a review of the state of current clinical trials.* Stem cells international, 2010. **2010**: p. 369578. DOI: 10.4061/2010/369578 Cited on page(s) 43

[251] Joyce, N., et al., *Mesenchymal stem cells for the treatment of neurodegenerative disease.* Regenerative Medicine, 2010. **5**(6): p. 933–946. DOI: 10.2217/rme.10.72 Cited on page(s)

[252] Vaquero, J.V.J. and M. Zurita, *Functional recovery after severe CNS trauma: Current perspectives for cell therapy with bone marrow stromal cells.* Progress in Neurobiology, 2011. **93**(3): p. 341–349. DOI: 10.1016/j.pneurobio.2010.12.002 Cited on page(s) 43

[253] Crigler, L., et al., *Human mesenchymal stem cell subpopulations express a variety of neuro-regulatory molecules and promote neuronal cell survival and neuritogenesis.* Experimental Neurology, 2006. **198**(1): p. 54–64. DOI: 10.1016/j.expneurol.2005.10.029 Cited on page(s) 43

[254] Ankeny, D.P., D.M. McTigue, and L.B. Jakeman, *Bone marrow transplants provide tissue protection and directional guidance for axons after contusive spinal cord injury in rats.* Experimental Neurology, 2004. **190**(1): p. 17–31. DOI: 10.1016/j.expneurol.2004.05.045 Cited on page(s) 43

[255] Akiyama, Y., C. Radtke, and J.D. Kocsis, *Remyelination of the rat spinal cord by transplantation of identified bone marrow stromal cells.* Journal of Neuroscience, 2002. **22**(15): p. 6623–6630. Cited on page(s) 43

[256] Inoue, M., et al., *Comparative analysis of remyelinating potential of focal and intravenous administration of autologous bone marrow cells into the rat demyelinated spinal cord.* Glia, 2003. **44**(2): p. 111–118. DOI: 10.1002/glia.10285 Cited on page(s) 43

[257] Rivera, F.J., et al., *Mesenchymal Stem Cells Promote Oligodendroglial Differentiation in Hippocampal Slice Cultures.* Cellular Physiology and Biochemistry, 2009. **24**(3–4): p. 317–324. DOI: 10.1159/000233256 Cited on page(s) 44

[258] Li, L.A.L.L.A., et al., *Transplantation of Marrow Stromal Cells Restores Cerebral Blood Flow and Reduces Cerebral Atrophy in Rats with Traumatic Brain Injury: In vivo MRI Study.* Journal of Neurotrauma, 2011. **28**(4): p. 535–545. DOI: 10.1089/neu.2010.1619 Cited on page(s) 44

[259] Fernandes, K.J.L., et al., *A dermal niche for multipotent adult skin-derived precursor cells.* Nature Cell Biology, 2004. **6**(11): p. 1082–U16. DOI: 10.1038/ncb1181 Cited on page(s) 44

[260] Fernandes, K.J.L., J.G. Toma, and F.D. Miller, *Multipotent skin-derived precursors: adult neural crest-related precursors with therapeutic potential.* Philosophical Transactions of the Royal Society B-Biological Sciences, 2008. **363**(1489): p. 185–198. DOI: 10.1098/rstb.2006.2020 Cited on page(s)

[261] Toma, J.G., et al., *Isolation of multipotent adult stem cells from the dermis of mammalian skin.* Nature Cell Biology, 2001. **3**(9): p. 778–784. DOI: 10.1038/ncb0901-778 Cited on page(s) 44

[262] Toma, J.G., et al., *Isolation and characterization of multipotent skin-derived precursors from human skin.* Stem Cells, 2005. **23**(6): p. 727–737. DOI: 10.1634/stemcells.2004-0134 Cited on page(s) 44

[263] Hunt, D.P.J., C. Jahoda, and S. Chandran, *Multipotent skin-derived precursors: from biology to clinical translation.* Current Opinion in Biotechnology, 2009. **20**(5): p. 522–530. DOI: 10.1016/j.copbio.2009.10.004 Cited on page(s) 44

[264] Amoh, Y., et al., *Multipotent nestin-positive, keratin-negative hair-follicle bulge stem cells can form neurons.* Proceedings of the National Academy of Sciences of the United States of America, 2005. **102**(15): p. 5530–5534. DOI: 10.1073/pnas.0501263102 Cited on page(s) 44

[265] Fernandes, K.J.L., et al., *Analysis of the neurogenic potential of multipotent skin-derived precursors.* Experimental Neurology, 2006. **201**(1): p. 32–48. DOI: 10.1016/j.expneurol.2006.03.018 Cited on page(s) 44

[266] Uchugonova, A., et al., *The Bulge Area Is the Origin of Nestin-Expressing Pluripotent Stem Cells of the Hair Follicle.* Journal of Cellular Biochemistry, 2011. **112**(8): p. 2046–2050. DOI: 10.1002/jcb.23122 Cited on page(s) 44

[267] Amoh, Y., et al., *Multipotent hair follicle stem cells promote repair of spinal cord injury and recovery of walking function.* Cell Cycle, 2008. **7**(12): p. 1865–1869. DOI: 10.4161/cc.7.12.6056 Cited on page(s) 44, 45

[268] Biernaskie, J., et al., *Skin-derived precursors generate myelinating Schwann cells that promote remyelination and functional recovery after contusion spinal cord injury.* Journal of Neuroscience, 2007. **27**(36): p. 9545–9559. DOI: 10.1523/JNEUROSCI.1930-07.2007 Cited on page(s) 44, 46

[269] Biernaskie, J.A., et al., *Isolation of skin-derived precursors (SKPs) and differentiation and enrichment of their Schwann cell progeny.* Nature Protocols, 2006. **1**(6): p. 2803–2812. DOI: 10.1038/nprot.2006.422 Cited on page(s) 44

[270] Gago, N., et al., *Age-Dependent Depletion of Human Skin-Derived Progenitor Cells.* Stem Cells, 2009. **27**(5): p. 1164–1172. DOI: 10.1002/stem.27 Cited on page(s) 46

[271] Szebenyi, G., et al., *Interstitial branches develop from active regions of the axon demarcated by the primary growth cone during pausing behaviors.* J Neurosci, 1998. **18**(19): p. 7930–40. Cited on page(s) 47

[272] Dotti, C.G., C.A. Sullivan, and G.A. Banker, *The establishment of polarity by hippocampal neurons in culture.* J Neurosci, 1988. **8**(4): p. 1454–68. Cited on page(s) 47

[273] Wen, Z. and J.Q. Zheng, *Directional guidance of nerve growth cones.* Curr Opin Neurobiol, 2006. **16**(1): p. 52–8. DOI: 10.1016/j.conb.2005.12.005 Cited on page(s) 47

[274] Farrar, N.R. and G.E. Spencer, *Pursuing a 'turning point' in growth cone research.* Dev Biol, 2008. **318**(1): p. 102–11. DOI: 10.1016/j.ydbio.2008.03.012 Cited on page(s) 47

[275] Kalil, K., G. Szebenyi, and E.W. Dent, *Common mechanisms underlying growth cone guidance and axon branching.* J Neurobiol, 2000. **44**(2): p. 145–58. DOI: 10.1002/1097-4695(200008)44:2%3C145::AID-NEU5%3E3.0.CO;2-X Cited on page(s) 47

[276] Suter, D.M. and K.E. Miller, *The emerging role of forces in axonal elongation.* Progress in Neurobiology, 2011. **94**(2): p. 91–101. DOI: 10.1016/j.pneurobio.2011.04.002 Cited on page(s) 48

[277] Flanagan, L.A., et al., *Neurite branching on deformable substrates.* Neuroreport, 2002. **13**(18): p. 2411–2415. DOI: 10.1097/00001756-200212200-00007 Cited on page(s) 48

[278] Gunn, J.W., S.D. Turner, and B.K. Mann, *Adhesive and mechanical properties of hydrogels influence neurite extension.* Journal of Biomedical Materials Research Part A, 2005. **72A**(1): p. 91–97. DOI: 10.1002/jbm.a.30203 Cited on page(s) 48

[279] Marquardt, L. and R.K. Willits, *Neurite growth in PEG gels: Effect of mechanical stiffness and laminin concentration.* Journal of Biomedical Materials Research Part A, 2011. **98A**(1): p. 1–6. DOI: 10.1002/jbm.a.33044 Cited on page(s) 48

[280] Sundararaghavan, H.G., et al., *Neurite Growth in 3D Collagen Gels With Gradients of Mechanical Properties.* Biotechnology and Bioengineering, 2009. **102**(2): p. 632–643. DOI: 10.1002/bit.22074 Cited on page(s) 48

[281] Willits, R.K. and S.L. Skornia, *Effect of collagen gel stiffness on neurite extension.* Journal of Biomaterials Science-Polymer Edition, 2004. **15**(12): p. 1521–1531. DOI: 10.1163/1568562042459698 Cited on page(s) 48

[282] Leach, J.B., et al., *Neurite outgrowth and branching of PC12 cells on very soft substrates sharply decreases below a threshold of substrate rigidity.* Journal of Neural Engineering, 2007. **4**(2): p. 26–34. DOI: 10.1088/1741-2560/4/2/003 Cited on page(s) 48

[283] Pfister, B.J., et al., *Extreme stretch growth of integrated axons.* Journal of Neuroscience, 2004. **24**(36): p. 7978–7983. DOI: 10.1523/JNEUROSCI.1974-04.2004 Cited on page(s) 48

[284] Pfister, B.J., et al., *Development of transplantable nervous tissue constructs comprised of stretch-grown axons.* Journal of Neuroscience Methods, 2006. **153**(1): p. 95–103. DOI: 10.1016/j.jneumeth.2005.10.012 Cited on page(s)

[285] Smith, D.H., J.A. Wolf, and D.F. Meaney, *A new strategy to produce sustained growth of central nervous system axons: Continuous mechanical tension.* Tissue Engineering, 2001. **7**(2): p. 131–139. DOI: 10.1089/107632701300062714 Cited on page(s) 48

[286] Bray, D., *Axonal growth in response to experimentally applied mechanical tension.* Developmental Biology, 1984. **102**(2): p. 379–389. DOI: 10.1016/0012-1606(84)90202-1 Cited on page(s) 48

[287] Bueno, F.R. and S.B. Shah, *Implications of tensile loading for the tissue engineering of nerves.* Tissue Engineering Part B-Reviews, 2008. **14**(3): p. 219–233. DOI: 10.1089/ten.teb.2008.0020 Cited on page(s)

[288] Lamoureux, P., et al., *Growth and Elongation Within and Along the Axon.* Developmental Neurobiology, 2010. **70**(3): p. 135–149. DOI: 10.1002/dneu.20764 Cited on page(s) 48

[289] Pfister, B.J., et al., *Stretch-grown axons retain the ability to transmit active electrical signals.* Febs Letters, 2006. **580**(14): p. 3525–3531. DOI: 10.1016/j.febslet.2006.05.030 Cited on page(s) 48, 49

[290] Pfister, B.J., et al., *Neural engineering to produce in vitro nerve constructs and neurointerface.* Neurosurgery, 2007. **60**(1): p. 137–141. DOI: 10.1227/01.NEU.0000249197.61280.1D Cited on page(s)

[291] Pfister, B.P., et al., *Engineering nerve constructs for clinical application.* Journal of Neurotrauma, 2004. **21**(9): p. P98. Cited on page(s) 48

[292] Huang, J.H., et al., *Long-Term Survival and Integration of Transplanted Engineered Nervous Tissue Constructs Promotes Peripheral Nerve Regeneration.* Tissue Engineering Part A, 2009. **15**(7): p. 1677–1685. DOI: 10.1089/ten.tea.2008.0294 Cited on page(s) 49

[293] Iwata, A., et al., *Long-term survival and outgrowth of mechanically engineered nervous tissue constructs implanted into spinal cord lesions.* Tissue Engineering, 2006. **12**(1): p. 101–110. DOI: 10.1089/ten.2006.12.101 Cited on page(s) 49

[294] Hoffman-Kim, D., J.A. Mitchel, and R.V. Bellamkonda, *Topography, Cell Response, and Nerve Regeneration*, in *Annual Review of Biomedical Engineering, Vol 12*, M.L.D.J.S.G.M.L. Yarmush, Editor. 2010. p. 203–231. Cited on page(s) 49, 50

[295] Khan, S. and G. Newaz, *A comprehensive review of surface modification for neural cell adhesion and patterning.* Journal of Biomedical Materials Research Part A, 2010. **93A**(3): p. 1209–1224. DOI: 10.1002/jbm.a.32698 Cited on page(s)

[296] Roach, P., et al., *Surface strategies for control of neuronal cell adhesion: A review.* Surface Science Reports, 2010. **65**(6): p. 145–173. DOI: 10.1016/j.surfrep.2010.07.001 Cited on page(s) 49, 64

[297] Boland, T., et al., *Application of inkjet printing to tissue engineering.* Biotechnology Journal, 2006. **1**(9): p. 910–917. DOI: 10.1002/biot.200600081 Cited on page(s) 51

[298] Fedorovich, N.E., et al., *Hydrogels as extracellular matrices for skeletal tissue engineering: state-of-the-art and novel application in organ printing.* Tissue Engineering, 2007. **13**(8): p. 1905–1925. DOI: 10.1089/ten.2006.0175 Cited on page(s)

[299] Nakamura, M., et al., *Biocompatible inkjet printing technique for designed seeding of individual living cells.* Tissue Engineering, 2005. **11**(11–12): p. 1658–1666. DOI: 10.1089/ten.2005.11.1658 Cited on page(s)

[300] Ringeisen, B.R., et al., *Jet-based methods to print living cells.* Biotechnology Journal, 2006. **1**(9): p. 930–948. DOI: 10.1002/biot.200600058 Cited on page(s)

[301] Roth, E.A., et al., *Inkjet printing for high-throughput cell patterning*. Biomaterials, 2004. **25**(17): p. 3707–3715. DOI: 10.1016/j.biomaterials.2003.10.052 Cited on page(s)

[302] Yamazoe, H. and T. Tanabe, *Cell micropatterning on an albumin-based substrate using an inkjet printing technique*. Journal of Biomedical Materials Research Part A, 2009. **91A**(4): p. 1202–1209. DOI: 10.1002/jbm.a.32312 Cited on page(s) 51

[303] Kofron, C.M., et al., *Neurite Outgrowth at the Biomimetic Interface*. Annals of Biomedical Engineering, 2010. **38**(6): p. 2210–2225. DOI: 10.1007/s10439-010-0054-y Cited on page(s) 51, 52

[304] Miller, C., S. Jeftinija, and S. Mallapragada, *Micropatterned Schwann cell-seeded biodegradable polymer substrates significantly enhance neurite alignment and outgrowth*. Tissue Engineering, 2001. **7**(6): p. 705–715. DOI: 10.1089/107632701753337663 Cited on page(s)

[305] Peerani, R., et al., *Patterning Mouse and Human Embryonic Stem Cells Using Micro-contact Printing*, in *Methods in Molecular Biology*, J.S.W.L. Audet, Editor. 2009. p. 21–33. Cited on page(s) 51

[306] Arumuganathar, S., et al., *A novel direct aerodynamically assisted threading methodology for generating biologically viable microthreads encapsulating living primary cells*. Journal of Applied Polymer Science, 2008. **107**(2): p. 1215–1225. DOI: 10.1002/app.27190 Cited on page(s) 51

[307] Townsend-Nicholson, A. and S.N. Jayasinghe, *Cell electrospinning: a unique biotechnique for encapsulating living organisms for generating active biological microthreads/scaffolds*. Biomacromolecules, 2006. **7**(12): p. 3364–3369. DOI: 10.1021/bm060649h Cited on page(s) 51

[308] Gomez, N., et al., *Immobilized nerve growth factor and microtopography have distinct effects on polarization versus axon elongation in hippocampal cells in culture*. Biomaterials, 2007. **28**(2): p. 271–284. DOI: 10.1016/j.biomaterials.2006.07.043 Cited on page(s) 51

[309] Jang, M.J., et al., *Directional neurite growth using carbon nanotube patterned substrates as a biomimetic cue*. Nanotechnology, 2010. **21**(23). DOI: 10.1088/0957-4484/21/23/235102 Cited on page(s)

[310] Li, J.M., H. McNally, and R. Shi, *Enhanced neurite alignment on micro-patterned poly-L-lactic acid films*. Journal of Biomedical Materials Research Part A, 2008. **87A**(2): p. 392–404. DOI: 10.1002/jbm.a.31814 Cited on page(s) 51

[311] Goldner, J.S., et al., *Neurite bridging across micropatterned grooves*. Biomaterials, 2006. **27**(3): p. 460–472. DOI: 10.1016/j.biomaterials.2005.06.035 Cited on page(s) 51

[312] Miller, C., S. Jeftinija, and S. Mallapragada, *Synergistic effects of physical and chemical guidance cues on neurite alignment and outgrowth on biodegradable polymer substrates*. Tissue Engineering, 2002. **8**(3): p. 367–378. DOI: 10.1089/107632702760184646 Cited on page(s) 51

[313] Yu, T.T. and M.S. Shoichet, *Guided cell adhesion and outgrowth in peptide-modified channels for neural tissue engineering.* Biomaterials, 2005. **26**(13): p. 1507–14. DOI: 10.1016/j.biomaterials.2004.05.012 Cited on page(s) 52

[314] Midha, R., et al., *Growth factor enhancement of peripheral nerve regeneration through a novel synthetic hydrogel tube.* J Neurosurg, 2003. **99**(3): p. 555–65. DOI: 10.3171/jns.2003.99.3.0555 Cited on page(s) 52

[315] Xu, X.M., et al., *Bridging Schwann cell transplants promote axonal regeneration from both the rostral and caudal stumps of transected adult rat spinal cord.* J Neurocytol, 1997. **26**(1): p. 1–16. DOI: 10.1023/A:1018557923309 Cited on page(s) 52

[316] Yu, M., et al., *Semiconductor nanomembrane tubes: three-dimensional confinement for controlled neurite outgrowth.* ACS Nano, 2011. **5**(4): p. 2447–57. DOI: 10.1021/nn103618d Cited on page(s) 52

[317] Elsdale, T. and J. Bard, *Collagen substrata for studies on cell behavior.* J Cell Biol, 1972. **54**(3): p. 626–37. DOI: 10.1083/jcb.54.3.626 Cited on page(s) 52

[318] Barnes, C.P., et al., *Nanofiber technology: designing the next generation of tissue engineering scaffolds.* Adv Drug Deliv Rev, 2007. **59**(14): p. 1413–33. DOI: 10.1016/j.addr.2007.04.022 Cited on page(s) 52

[319] Smith, L.A. and P.X. Ma, *Nano-fibrous scaffolds for tissue engineering.* Colloids Surf B Biointerfaces, 2004. **39**(3): p. 125–31. DOI: 10.1016/j.colsurfb.2003.12.004 Cited on page(s) 52

[320] Yang, F., et al., *Electrospinning of nano/micro scale poly(L-lactic acid) aligned fibers and their potential in neural tissue engineering.* Biomaterials, 2005. **26**(15): p. 2603–10. DOI: 10.1016/j.biomaterials.2004.06.051 Cited on page(s) 52

[321] Corey, J.M., et al., *Aligned electrospun nanofibers specify the direction of dorsal root ganglia neurite growth.* J Biomed Mater Res A, 2007. **83**(3): p. 636–45. DOI: 10.1002/jbm.a.31285 Cited on page(s) 52

[322] Shaw, D. and M.S. Shoichet, *Toward spinal cord injury repair strategies: peptide surface modification of expanded poly(tetrafluoroethylene) fibers for guided neurite outgrowth in vitro.* J Craniofac Surg, 2003. **14**(3): p. 308–16. DOI: 10.1097/00001665-200305000-00008 Cited on page(s) 52

[323] Wen, X. and P.A. Tresco, *Effect of filament diameter and extracellular matrix molecule precoating on neurite outgrowth and Schwann cell behavior on multifilament entubulation bridging device in vitro.* J Biomed Mater Res A, 2006. **76**(3): p. 626–37. DOI: 10.1002/jbm.a.30520 Cited on page(s) 52

[324] Karlsson, M., F. Johansson, and M. Kanje, *Polystyrene replicas of neuronal basal lamina act as excellent guides for regenerating neurites.* Acta Biomaterialia, 2011. **7**(7): p. 2910–2918. DOI: 10.1016/j.actbio.2011.03.029 Cited on page(s) 52

[325] Wang, D.Y., et al., *Microcontact printing of laminin on oxygen plasma activated substrates for the alignment and growth of Schwann cells.* J Biomed Mater Res B Appl Biomater, 2007. **80**(2): p. 447–53. DOI: 10.1002/jbm.b.30616 Cited on page(s) 52

[326] Schmalenberg, K.E. and K.E. Uhrich, *Micropatterned polymer substrates control alignment of proliferating Schwann cells to direct neuronal regeneration.* Biomaterials, 2005. **26**(12): p. 1423–30. DOI: 10.1016/j.biomaterials.2004.04.046 Cited on page(s) 52

[327] Kofron, C.M. and D. Hoffman-Kim, *Optimization by Response Surface Methodology of Confluent and Aligned Cellular Monolayers for Nerve Guidance.* Cell Mol Bioeng, 2009. **2**(4): p. 554–572. DOI: 10.1007/s12195-009-0087-1 Cited on page(s) 52

[328] Miner, J.H. and P.D. Yurchenco, *Laminin functions in tissue morphogenesis.* Annu Rev Cell Dev Biol, 2004. **20**: p. 255–84. DOI: 10.1146/annurev.cellbio.20.010403.094555 Cited on page(s) 54

[329] Selak, I., J.M. Foidart, and G. Moonen, *Laminin promotes cerebellar granule cells migration in vitro and is synthesized by cultured astrocytes.* Dev Neurosci, 1985. **7**(5–6): p. 278–85. DOI: 10.1159/000112296 Cited on page(s) 54

[330] Liesi, P., et al., *Domain-specific antibodies against the B2 chain of laminin inhibit neuronal migration in the neonatal rat cerebellum.* J Neurosci Res, 1995. **40**(2): p. 199–206. DOI: 10.1002/jnr.490400208 Cited on page(s) 54

[331] Smalheiser, N.R. and N.B. Schwartz, *Cranin: a laminin-binding protein of cell membranes.* Proc Natl Acad Sci U S A, 1987. **84**(18): p. 6457–61. DOI: 10.1073/pnas.84.18.6457 Cited on page(s) 54

[332] Smalheiser, N.R., *Cranin interacts specifically with the sulfatide-binding domain of laminin.* J Neurosci Res, 1993. **36**(5): p. 528–38. DOI: 10.1002/jnr.490360505 Cited on page(s) 54

[333] Goh, E.L., et al., *beta1-integrin mediates myelin-associated glycoprotein signaling in neuronal growth cones.* Mol Brain, 2008. **1**: p. 10. DOI: 10.1186/1756-6606-1-10 Cited on page(s) 54

[334] Plantman, S., et al., *Integrin-laminin interactions controlling neurite outgrowth from adult DRG neurons in vitro.* Mol Cell Neurosci, 2008. **39**(1): p. 50–62. DOI: 10.1016/j.mcn.2008.05.015 Cited on page(s) 54

[335] Kim, S.H., J. Turnbull, and S. Guimond, *Extracellular matrix and cell signalling: the dynamic cooperation of integrin, proteoglycan and growth factor receptor.* J Endocrinol, 2011. **209**(2): p. 139–51. DOI: 10.1530/JOE-10-0377 Cited on page(s) 54

[336] Wojcik-Stanaszek, L., A. Gregor, and T. Zalewska, *Regulation of neurogenesis by extracellular matrix and integrins.* Acta Neurobiol Exp (Wars), 2011. **71**(1): p. 103–12. Cited on page(s)

[337] Luo, B.H., C.V. Carman, and T.A. Springer, *Structural basis of integrin regulation and signaling.* Annu Rev Immunol, 2007. **25**: p. 619–47. DOI: 10.1146/annurev.immunol.25.022106.141618 Cited on page(s) 54

[338] Jacques, T.S., et al., *Neural precursor cell chain migration and division are regulated through different beta1 integrins.* Development, 1998. **125**(16): p. 3167–77. Cited on page(s) 54

[339] Li, S., et al., *Matrix assembly, regulation, and survival functions of laminin and its receptors in embryonic stem cell differentiation.* J Cell Biol, 2002. **157**(7): p. 1279–90. DOI: 10.1083/jcb.200203073 Cited on page(s) 54

[340] Varnum-Finney, B. and L.F. Reichardt, *Vinculin-deficient PC12 cell lines extend unstable lamellipodia and filopodia and have a reduced rate of neurite outgrowth.* J Cell Biol, 1994. **127**(4): p. 1071–84. DOI: 10.1083/jcb.127.4.1071 Cited on page(s) 54

[341] Jay, D.G., *A Src-astic response to mounting tension.* J Cell Biol, 2001. **155**(3): p. 327–30. DOI: 10.1083/jcb.200110019 Cited on page(s)

[342] Woo, S. and T.M. Gomez, *Rac1 and RhoA promote neurite outgrowth through formation and stabilization of growth cone point contacts.* J Neurosci, 2006. **26**(5): p. 1418–28. DOI: 10.1523/JNEUROSCI.4209-05.2006 Cited on page(s) 54

[343] Letourneau, P.C., M.L. Condic, and D.M. Snow, *Interactions of developing neurons with the extracellular matrix.* J Neurosci, 1994. **14**(3 Pt 1): p. 915–28. Cited on page(s) 54

[344] Reichardt, L.F. and K.J. Tomaselli, *Extracellular matrix molecules and their receptors: functions in neural development.* Annu Rev Neurosci, 1991. **14**: p. 531–70. DOI: 10.1146/annurev.ne.14.030191.002531 Cited on page(s) 54

[345] Wang, L. and J.L. Denburg, *A role for proteoglycans in the guidance of a subset of pioneer axons in cultured embryos of the cockroach.* Neuron, 1992. **8**(4): p. 701–14. DOI: 10.1016/0896-6273(92)90091-Q Cited on page(s) 54

[346] Walz, A., et al., *Essential role of heparan sulfates in axon navigation and targeting in the developing visual system.* Development, 1997. **124**(12): p. 2421–30. Cited on page(s)

[347] Irie, A., et al., *Specific heparan sulfate structures involved in retinal axon targeting.* Development, 2002. **129**(1): p. 61–70. Cited on page(s) 54

[348] de Wit, J. and J. Verhaagen, *Proteoglycans as modulators of axon guidance cue function.* Semi-aphorins: Receptor and Intracellular Signaling Mechanisms, 2007. **600**: p. 73–89. DOI: 10.1007/978-0-387-70956-7_7 Cited on page(s) 54

[349] Tom, V.J., et al., *Studies on the development and behavior of the dystrophic growth cone, the hallmark of regeneration failure, in an in vitro model of the glial scar and after spinal cord injury.* J Neurosci, 2004. **24**(29): p. 6531–9. DOI: 10.1523/JNEUROSCI.0994-04.2004 Cited on page(s) 54

[350] Kantor, D.B., et al., *Semaphorin 5A is a bifunctional axon guidance cue regulated by heparan and chondroitin sulfate proteoglycans.* Neuron, 2004. **44**(6): p. 961–75. DOI: 10.1016/j.neuron.2004.12.002 Cited on page(s) 54

[351] Li, G.N., J. Liu, and D. Hoffman-Kim, *Multi-molecular gradients of permissive and inhibitory cues direct neurite outgrowth.* Annals of Biomedical Engineering, 2008. **36**(6): p. 889–904. DOI: 10.1007/s10439-008-9486-z Cited on page(s) 54

[352] Jones, F.S. and P.L. Jones, *The tenascin family of ECM glycoproteins: structure, function, and regulation during embryonic development and tissue remodeling.* Dev Dyn, 2000. **218**(2): p. 235–59. DOI: 10.1002/(SICI)1097-0177(200006)218:2%3C235::AID-DVDY2%3E3.0.CO;2-G Cited on page(s) 55

[353] Lang, D.M., et al., *Tenascin-R and axon growth-promoting molecules are up-regulated in the regenerating visual pathway of the lizard (Gallotia galloti).* Dev Neurobiol, 2008. **68**(7): p. 899–916. DOI: 10.1002/dneu.20624 Cited on page(s) 55

[354] Lykissas, M.G., et al., *The role of neurotrophins in axonal growth, guidance, and regeneration.* Current Neurovascular Research, 2007. **4**(2): p. 143–151. DOI: 10.2174/156720207780637216 Cited on page(s) 55

[355] TessierLavigne, M. and C.S. Goodman, *The molecular biology of axon guidance.* Science, 1996. **274**(5290): p. 1123–1133. DOI: 10.1126/science.274.5290.1123 Cited on page(s)

[356] Chilton, J.K., *Molecular mechanisms of axon guidance.* Dev Biol, 2006. **292**(1): p. 13–24. DOI: 10.1016/j.ydbio.2005.12.048 Cited on page(s) 55

[357] Huang, E.J. and L.F. Reichardt, *Neurotrophins: roles in neuronal development and function.* Annu Rev Neurosci, 2001. **24**: p. 677–736. DOI: 10.1146/annurev.neuro.24.1.677 Cited on page(s) 55

[358] Hallbook, F., *Evolution of the vertebrate neurotrophin and Trk receptor gene families.* Curr Opin Neurobiol, 1999. **9**(5): p. 616–21. DOI: 10.1016/S0959-4388(99)00011-2 Cited on page(s) 55

[359] Levi-Montalcini, R., *The nerve growth factor: thirty-five years later.* Biosci Rep, 1987. **7**(9): p. 681–99. DOI: 10.1007/BF01116861 Cited on page(s) 55

[360] Cordon-Cardo, C., et al., *The trk tyrosine protein kinase mediates the mitogenic properties of nerve growth factor and neurotrophin-3.* Cell, 1991. **66**(1): p. 173–83. DOI: 10.1016/0092-8674(91)90149-S Cited on page(s) 55

[361] Kaplan, D.R., et al., *The trk proto-oncogene product: a signal transducing receptor for nerve growth factor.* Science, 1991. **252**(5005): p. 554–8. DOI: 10.1126/science.1850549 Cited on page(s)

[362] Klein, R., et al., *The trk proto-oncogene encodes a receptor for nerve growth factor.* Cell, 1991. **65**(1): p. 189–97. DOI: 10.1016/0092-8674(91)90419-Y Cited on page(s) 55

[363] Berkemeier, L.R., et al., *Neurotrophin-5: a novel neurotrophic factor that activates trk and trkB.* Neuron, 1991. **7**(5): p. 857–66. DOI: 10.1016/0896-6273(91)90287-A Cited on page(s) 55, 56

[364] Ip, N.Y., et al., *Mammalian neurotrophin-4: structure, chromosomal localization, tissue distribution, and receptor specificity.* Proc Natl Acad Sci U S A, 1992. **89**(7): p. 3060–4. DOI: 10.1073/pnas.89.7.3060 Cited on page(s) 55

[365] Lamballe, F., R. Klein, and M. Barbacid, *trkC, a new member of the trk family of tyrosine protein kinases, is a receptor for neurotrophin-3.* Cell, 1991. **66**(5): p. 967–79. DOI: 10.1016/0092-8674(91)90442-2 Cited on page(s) 55

[366] Huang, E.J., et al., *Expression of Trk receptors in the developing mouse trigeminal ganglion: in vivo evidence for NT-3 activation of TrkA and TrkB in addition to TrkC.* Development, 1999. **126**(10): p. 2191–203. Cited on page(s) 55

[367] Rodriguez-Tebar, A., G. Dechant, and Y.A. Barde, *Neurotrophins: structural relatedness and receptor interactions.* Philos Trans R Soc Lond B Biol Sci, 1991. **331**(1261): p. 255–8. DOI: 10.1098/rstb.1991.0013 Cited on page(s) 55

[368] Barde, Y.A., D. Edgar, and H. Thoenen, *Purification of a new neurotrophic factor from mammalian brain.* EMBO J, 1982. **1**(5): p. 549–53. Cited on page(s) 55

[369] Leibrock, J., et al., *Molecular cloning and expression of brain-derived neurotrophic factor.* Nature, 1989. **341**(6238): p. 149–52. DOI: 10.1038/341149a0 Cited on page(s) 55

[370] Carvalho, A.L., et al., *Role of the brain-derived neurotrophic factor at glutamatergic synapses.* Br J Pharmacol, 2007. DOI: 10.1038/sj.bjp.0707509 Cited on page(s) 55

[371] Hohn, A., et al., *Identification and characterization of a novel member of the nerve growth factor/brain-derived neurotrophic factor family.* Nature, 1990. **344**(6264): p. 339–41. DOI: 10.1038/344339a0 Cited on page(s) 55

[372] Maisonpierre, P.C., et al., *Neurotrophin-3: a neurotrophic factor related to NGF and BDNF.* Science, 1990. **247**(4949 Pt 1): p. 1446–51. DOI: 10.1126/science.2321006 Cited on page(s)

[373] Rosenthal, A., et al., *Primary structure and biological activity of a novel human neurotrophic factor.* Neuron, 1990. **4**(5): p. 767–73. DOI: 10.1016/0896-6273(90)90203-R Cited on page(s) 55

[374] Tobias, C.A., et al., *Delayed grafting of BDNF and NT-3 producing fibroblasts into the injured spinal cord stimulates sprouting, partially rescues axotomized red nucleus neurons from loss and atrophy, and provides limited regeneration.* Exp Neurol, 2003. **184**(1): p. 97–113. DOI: 10.1016/S0014-4886(03)00394-7 Cited on page(s) 55

[375] Yuan, X.B., et al., *Signalling and crosstalk of Rho GTPases in mediating axon guidance.* Nat Cell Biol, 2003. **5**(1): p. 38–45. DOI: 10.1038/ncb895 Cited on page(s) 55

[376] Tessier-Lavigne, M. and C.S. Goodman, *The molecular biology of axon guidance.* Science, 1996. **274**(5290): p. 1123–33. DOI: 10.1126/science.274.5290.1123 Cited on page(s) 55

[377] Hallbook, F., C.F. Ibanez, and H. Persson, *Evolutionary studies of the nerve growth factor family reveal a novel member abundantly expressed in Xenopus ovary.* Neuron, 1991. **6**(5): p. 845–58. DOI: 10.1016/0896-6273(91)90180-8 Cited on page(s) 56

[378] Lewin, G.R. and Y.A. Barde, *Physiology of the neurotrophins.* Annu Rev Neurosci, 1996. **19**: p. 289–317. DOI: 10.1146/annurev.ne.19.030196.001445 Cited on page(s) 56

[379] Lykissas, M.G., et al., *The role of neurotrophins in axonal growth, guidance, and regeneration.* Curr Neurovasc Res, 2007. **4**(2): p. 143–51. DOI: 10.2174/156720207780637216 Cited on page(s) 56

[380] Blesch, A. and M.H. Tuszynski, *Cellular GDNF delivery promotes growth of motor and dorsal column sensory axons after partial and complete spinal cord transections and induces remyelination.* J Comp Neurol, 2003. **467**(3): p. 403–17. DOI: 10.1002/cne.10934 Cited on page(s) 56

[381] Jungnickel, J., et al., *Faster nerve regeneration after sciatic nerve injury in mice over-expressing basic fibroblast growth factor.* J Neurobiol, 2006. **66**(9): p. 940–8. DOI: 10.1002/neu.20265 Cited on page(s) 56

[382] Sendtner, M., et al., *Endogenous ciliary neurotrophic factor is a lesion factor for axotomized motoneurons in adult mice.* J Neurosci, 1997. **17**(18): p. 6999–7006. Cited on page(s) 56

[383] Kelleher, M.O., et al., *The use of ciliary neurotrophic factor to promote recovery after peripheral nerve injury by delivering it at the site of the cell body.* Acta Neurochir (Wien), 2006. **148**(1): p. 55–60; discussion 60-1. DOI: 10.1007/s00701-005-0631-2 Cited on page(s) 56

[384] McCallister, W.V., et al., *Regeneration along intact nerves using nerve growth factor and ciliary neurotrophic factor.* J Reconstr Microsurg, 2004. **20**(6): p. 473–81. DOI: 10.1055/s-2004-833499 Cited on page(s) 56

[385] *Unified nomenclature for Eph family receptors and their ligands, the ephrins. Eph Nomenclature Committee.* Cell, 1997. **90**(3): p. 403–4. DOI: 10.1016/S0092-8674(00)80500-0 Cited on page(s) 56

[386] Gale, N.W., et al., *Eph receptors and ligands comprise two major specificity subclasses and are reciprocally compartmentalized during embryogenesis.* Neuron, 1996. **17**(1): p. 9–19. DOI: 10.1016/S0896-6273(00)80276-7 Cited on page(s) 56

[387] Himanen, J.P. and D.B. Nikolov, *Eph signaling: a structural view.* Trends Neurosci, 2003. **26**(1): p. 46–51. DOI: 10.1016/S0166-2236(02)00005-X Cited on page(s) 56

[388] Hattori, M., M. Osterfield, and J.G. Flanagan, *Regulated cleavage of a contact-mediated axon repellent.* Science, 2000. **289**(5483): p. 1360–5. DOI: 10.1126/science.289.5483.1360 Cited on page(s) 56

[389] Symonds, A.C., et al., *Reinnervation of the superior colliculus delays down-regulation of ephrin A2 in neonatal rat.* Exp Neurol, 2001. **170**(2): p. 364–70. DOI: 10.1006/exnr.2001.7722 Cited on page(s) 56

[390] Mellitzer, G., Q. Xu, and D.G. Wilkinson, *Control of cell behaviour by signalling through Eph receptors and ephrins.* Curr Opin Neurobiol, 2000. **10**(3): p. 400–8. DOI: 10.1016/S0959-4388(00)00095-7 Cited on page(s) 56

[391] Orioli, D. and R. Klein, *The Eph receptor family: axonal guidance by contact repulsion.* Trends Genet, 1997. **13**(9): p. 354–9. DOI: 10.1016/S0168-9525(97)01220-1 Cited on page(s) 56

[392] Cruz-Orengo, L., et al., *Reduction of EphA4 receptor expression after spinal cord injury does not induce axonal regeneration or return of tcMMEP response.* Neurosci Lett, 2007. **418**(1): p. 49–54. DOI: 10.1016/j.neulet.2007.03.015 Cited on page(s) 56

[393] Klein, R., *Cell sorting during regenerative tissue formation.* Cell, 2010. **143**(1): p. 32–4. DOI: 10.1016/j.cell.2010.09.018 Cited on page(s) 56

[394] Parrinello, S., et al., *EphB signaling directs peripheral nerve regeneration through Sox2-dependent Schwann cell sorting.* Cell, 2010. **143**(1): p. 145–55. DOI: 10.1016/j.cell.2010.08.039 Cited on page(s) 56

[395] Palmer, A. and R. Klein, *Multiple roles of ephrins in morphogenesis, neuronal networking, and brain function.* Genes Dev, 2003. **17**(12): p. 1429–50. DOI: 10.1101/gad.1093703 Cited on page(s) 56

[396] *Unified nomenclature for the semaphorins/collapsins. Semaphorin Nomenclature Committee.* Cell, 1999. **97**(5): p. 551–2. Cited on page(s) 56

[397] Luo, Y., D. Raible, and J.A. Raper, *Collapsin: a protein in brain that induces the collapse and paralysis of neuronal growth cones.* Cell, 1993. **75**(2): p. 217–27. DOI: 10.1016/0092-8674(93)80064-L Cited on page(s) 57

[398] Kumanogoh, A. and H. Kikutani, *Semaphorins and their receptors: novel features of neural guidance molecules.* Proc Jpn Acad Ser B Phys Biol Sci, 2010. **86**(6): p. 611–20. DOI: 10.2183/pjab.86.611 Cited on page(s) 57

[399] Raper, J.A. and J.P. Kapfhammer, *The enrichment of a neuronal growth cone collapsing activity from embryonic chick brain.* Neuron, 1990. **4**(1): p. 21–9. DOI: 10.1016/0896-6273(90)90440-Q Cited on page(s) 57

[400] Tran, T.S., A.L. Kolodkin, and R. Bharadwaj, *Semaphorin regulation of cellular morphology.* Annu Rev Cell Dev Biol, 2007. **23**: p. 263–92. DOI: 10.1146/annurev.cellbio.22.010605.093554 Cited on page(s) 57

[401] Fujisawa, H., *Discovery of semaphorin receptors, neuropilin and plexin, and their functions in neural development.* J Neurobiol, 2004. **59**(1): p. 24–33. DOI: 10.1002/neu.10337 Cited on page(s) 57

[402] Tamagnone, L., et al., *Plexins are a large family of receptors for transmembrane, secreted, and GPI-anchored semaphorins in vertebrates.* Cell, 1999. **99**(1): p. 71–80. DOI: 10.1016/S0092-8674(00)80063-X Cited on page(s)

[403] Liu, B.P. and S.M. Strittmatter, *Semaphorin-mediated axonal guidance via Rho-related G proteins.* Curr Opin Cell Biol, 2001. **13**(5): p. 619–26. DOI: 10.1016/S0955-0674(00)00260-X Cited on page(s) 57

[404] Nakamura, F., R.G. Kalb, and S.M. Strittmatter, *Molecular basis of semaphorin-mediated axon guidance.* J Neurobiol, 2000. **44**(2): p. 219–29. DOI: 10.1002/1097-4695(200008)44:2%3C219::AID-NEU11%3E3.0.CO;2-W Cited on page(s) 57

[405] de Wit, J. and J. Verhaagen, *Role of semaphorins in the adult nervous system.* Prog Neurobiol, 2003. **71**(2–3): p. 249–67. DOI: 10.1016/j.pneurobio.2003.06.001 Cited on page(s) 57

[406] Ara, J., et al., *Modulation of sciatic nerve expression of class 3 semaphorins by nerve injury.* Neurochem Res, 2004. **29**(6): p. 1153–9. DOI: 10.1023/B:NERE.0000023602.72354.82 Cited on page(s) 57

[407] De Winter, F., et al., *Injury-induced class 3 semaphorin expression in the rat spinal cord.* Exp Neurol, 2002. **175**(1): p. 61–75. DOI: 10.1006/exnr.2002.7884 Cited on page(s) 57

[408] Bannerman, P., et al., *Peripheral nerve regeneration is delayed in neuropilin 2-deficient mice.* J Neurosci Res, 2008. **86**(14): p. 3163–9. DOI: 10.1002/jnr.21766 Cited on page(s) 57

[409] Scarlato, M., et al., *Induction of neuropilins-1 and -2 and their ligands, Sema3A, Sema3F, and VEGF, during Wallerian degeneration in the peripheral nervous system.* Exp Neurol, 2003. **183**(2): p. 489–98. DOI: 10.1016/S0014-4886(03)00046-3 Cited on page(s) 57

[410] Koppel, A.M., et al., *A 70 amino acid region within the semaphorin domain activates specific cellular response of semaphorin family members.* Neuron, 1997. **19**(3): p. 531–7. DOI: 10.1016/S0896-6273(00)80369-4 Cited on page(s) 57

[411] Bagnard, D., et al., *Semaphorins act as attractive and repulsive guidance signals during the development of cortical projections.* Development, 1998. **125**(24): p. 5043–53. Cited on page(s) 57

[412] Itoh, A., et al., *Cloning and expressions of three mammalian homologues of Drosophila slit suggest possible roles for Slit in the formation and maintenance of the nervous system.* Brain Res Mol Brain Res, 1998. **62**(2): p. 175–86. DOI: 10.1016/S0169-328X(98)00224-1 Cited on page(s) 57

[413] Brose, K., et al., *Slit proteins bind Robo receptors and have an evolutionarily conserved role in repulsive axon guidance.* Cell, 1999. **96**(6): p. 795–806. DOI: 10.1016/S0092-8674(00)80590-5 Cited on page(s) 57

[414] Li, H.S., et al., *Vertebrate slit, a secreted ligand for the transmembrane protein roundabout, is a repellent for olfactory bulb axons.* Cell, 1999. **96**(6): p. 807–18. DOI: 10.1016/S0092-8674(00)80591-7 Cited on page(s) 57

[415] Huminiecki, L., et al., *Magic roundabout is a new member of the roundabout receptor family that is endothelial specific and expressed at sites of active angiogenesis.* Genomics, 2002. **79**(4): p. 547–52. DOI: 10.1006/geno.2002.6745 Cited on page(s) 57

[416] Kidd, T., et al., *Dosage-sensitive and complementary functions of roundabout and commissureless control axon crossing of the CNS midline.* Neuron, 1998. **20**(1): p. 25–33. DOI: 10.1016/S0896-6273(00)80431-6 Cited on page(s)

[417] Yuan, S.S., et al., *Cloning and functional studies of a novel gene aberrantly expressed in RB-deficient embryos.* Dev Biol, 1999. **207**(1): p. 62–75. DOI: 10.1006/dbio.1998.9141 Cited on page(s) 57

[418] Nguyen-Ba-Charvet, K.T. and A. Chedotal, *Role of Slit proteins in the vertebrate brain.* J Physiol Paris, 2002. **96**(1–2): p. 91–8. DOI: 10.1016/S0928-4257(01)00084-5 Cited on page(s) 57

[419] Nguyen Ba-Charvet, K.T., et al., *Diversity and specificity of actions of Slit2 proteolytic fragments in axon guidance.* J Neurosci, 2001. **21**(12): p. 4281–9. Cited on page(s) 57

[420] Long, H., et al., *Conserved roles for Slit and Robo proteins in midline commissural axon guidance.* Neuron, 2004. **42**(2): p. 213–23. DOI: 10.1016/S0896-6273(04)00179-5 Cited on page(s) 58

[421] Sabatier, C., et al., *The divergent Robo family protein rig-1/Robo3 is a negative regulator of slit responsiveness required for midline crossing by commissural axons.* Cell, 2004. **117**(2): p. 157–69. DOI: 10.1016/S0092-8674(04)00303-4 Cited on page(s) 58

[422] Andrews, W.D., M. Barber, and J.G. Parnavelas, *Slit-Robo interactions during cortical development.* J Anat, 2007. **211**(2): p. 188–98. DOI: 10.1111/j.1469-7580.2007.00750.x Cited on page(s) 58

[423] Marillat, V., et al., *Spatiotemporal expression patterns of slit and robo genes in the rat brain.* J Comp Neurol, 2002. **442**(2): p. 130–55. DOI: 10.1002/cne.10068 Cited on page(s)

[424] Dugan, J.P., et al., *Midbrain dopaminergic axons are guided longitudinally through the diencephalon by Slit/Robo signals.* Mol Cell Neurosci, 2011. **46**(1): p. 347–56. DOI: 10.1016/j.mcn.2010.11.003 Cited on page(s) 58

[425] Thompson, H., et al., *Robo2 is required for Slit-mediated intraretinal axon guidance.* Dev Biol, 2009. **335**(2): p. 418–26. DOI: 10.1016/j.ydbio.2009.09.034 Cited on page(s) 58

[426] Tayler, T.D., M.B. Robichaux, and P.A. Garrity, *Compartmentalization of visual centers in the Drosophila brain requires Slit and Robo proteins.* Development, 2004. **131**(23): p. 5935–45. DOI: 10.1242/dev.01465 Cited on page(s) 58

[427] Wehrle, R., et al., *Expression of netrin-1, slit-1 and slit-3 but not of slit-2 after cerebellar and spinal cord lesions.* Eur J Neurosci, 2005. **22**(9): p. 2134–44. DOI: 10.1111/j.1460-9568.2005.04419.x Cited on page(s) 58

[428] Hagino, S., et al., *Slit and glypican-1 mRNAs are coexpressed in the reactive astrocytes of the injured adult brain.* Glia, 2003. **42**(2): p. 130–8. DOI: 10.1002/glia.10207 Cited on page(s) 58

[429] Fujiwara, T., et al., *mRNA expression changes of slit proteins following peripheral nerve injury in the rat model.* J Chem Neuroanat, 2008. **36**(3–4): p. 170–6. DOI: 10.1016/j.jchemneu.2008.07.007 Cited on page(s) 58

[430] Bloechlinger, S., L.A. Karchewski, and C.J. Woolf, *Dynamic changes in glypican-1 expression in dorsal root ganglion neurons after peripheral and central axonal injury.* Eur J Neurosci, 2004. **19**(5): p. 1119–32. DOI: 10.1111/j.1460-9568.2004.03262.x Cited on page(s) 58

[431] Kennedy, T.E., et al., *Netrins are diffusible chemotropic factors for commissural axons in the embryonic spinal cord.* Cell, 1994. **78**(3): p. 425–35. DOI: 10.1016/0092-8674(94)90421-9 Cited on page(s) 58

[432] Serafini, T., et al., *The netrins define a family of axon outgrowth-promoting proteins homologous to C. elegans UNC-6.* Cell, 1994. **78**(3): p. 409–24. DOI: 10.1016/0092-8674(94)90420-0 Cited on page(s) 58

[433] Chan, S.S., et al., *UNC-40, a C. elegans homolog of DCC (Deleted in Colorectal Cancer), is required in motile cells responding to UNC-6 netrin cues.* Cell, 1996. **87**(2): p. 187–95. DOI: 10.1016/S0092-8674(00)81337-9 Cited on page(s) 58

[434] Keino-Masu, K., et al., *Deleted in Colorectal Cancer (DCC) encodes a netrin receptor.* Cell, 1996. **87**(2): p. 175–85. DOI: 10.1016/S0092-8674(00)81336-7 Cited on page(s)

[435] Vielmetter, J., et al., *Neogenin, an avian cell surface protein expressed during terminal neuronal differentiation, is closely related to the human tumor suppressor molecule deleted in colorectal cancer.* J Cell Biol, 1994. **127**(6 Pt 2): p. 2009–20. DOI: 10.1083/jcb.127.6.2009 Cited on page(s) 58

[436] Li, W., et al., *Activation of FAK and Src are receptor-proximal events required for netrin signaling.* Nat Neurosci, 2004. **7**(11): p. 1213–21. DOI: 10.1038/nn1329 Cited on page(s) 58

[437] De Vries, M. and H.M. Cooper, *Emerging roles for neogenin and its ligands in CNS development.* J Neurochem, 2008. **106**(4): p. 1483–92. DOI: 10.1111/j.1471-4159.2008.05485.x Cited on page(s) 58

[438] Hong, K., et al., *A ligand-gated association between cytoplasmic domains of UNC5 and DCC family receptors converts netrin-induced growth cone attraction to repulsion.* Cell, 1999. **97**(7): p. 927–41. DOI: 10.1016/S0092-8674(00)80804-1 Cited on page(s) 58

[439] Leonardo, E.D., et al., *Vertebrate homologues of C. elegans UNC-5 are candidate netrin receptors.* Nature, 1997. **386**(6627): p. 833–8. DOI: 10.1038/386833a0 Cited on page(s)

[440] Dillon, A.K., et al., *UNC5C is required for spinal accessory motor neuron development.* Mol Cell Neurosci, 2007. **35**(3): p. 482–9. DOI: 10.1016/j.mcn.2007.04.011 Cited on page(s) 58

[441] Stein, E. and M. Tessier-Lavigne, *Hierarchical organization of guidance receptors: silencing of netrin attraction by slit through a Robo/DCC receptor complex.* Science, 2001. **291**(5510): p. 1928–38. DOI: 10.1126/science.1058445 Cited on page(s) 58

[442] Cebria, F., et al., *Dissecting planarian central nervous system regeneration by the expression of neural-specific genes.* Dev Growth Differ, 2002. **44**(2): p. 135–46. DOI: 10.1046/j.1440-169x.2002.00629.x Cited on page(s) 58

[443] Ellezam, B., et al., *Expression of netrin-1 and its receptors DCC and UNC-5H2 after axotomy and during regeneration of adult rat retinal ganglion cells.* Exp Neurol, 2001. **168**(1): p. 105–15. DOI: 10.1006/exnr.2000.7589 Cited on page(s) 58

[444] Green, N.M., et al., Advances in Protein Chemistry, 1975: p. 85–133. Cited on page(s) 60

[445] Wegner, G.J., et al., *Fabrication of histidine-tagged fusion protein arrays for surface plasmon resonance imaging studies of protein-protein and protein-DNA interactions.* Anal Chem, 2003. **75**(18): p. 4740–6. DOI: 10.1021/ac0344438 Cited on page(s) 60

[446] Niemeyer, C.M., *The developments of semisynthetic DNA-protein conjugates.* Trends Biotechnol, 2002. **20**(9): p. 395–401. DOI: 10.1016/S0167-7799(02)02022-X Cited on page(s)

[447] Hendrickson, E.R., et al., *High sensitivity multianalyte immunoassay using covalent DNA-labeled antibodies and polymerase chain reaction.* Nucleic Acids Res, 1995. **23**(3): p. 522–9. DOI: 10.1093/nar/23.3.522 Cited on page(s) 60

[448] Rusmini, F., Z. Zhong, and J. Feijen, *Protein immobilization strategies for protein biochips.* Biomacromolecules, 2007. **8**(6): p. 1775–89. DOI: 10.1021/bm061197b Cited on page(s) 60

[449] Ganesan, R.e.a., *Multicomponent protein patterning of material surfaces.* Journal of Materials Chemistry, 2010. **20**: p. 7322–7331. DOI: 10.1039/b926690a Cited on page(s) 60

[450] Leipzig, N.D., et al., *Differentiation of neural stem cells in three-dimensional growth factor-immobilized chitosan hydrogel scaffolds.* Biomaterials, 2011. **32**(1): p. 57–64. DOI: 10.1016/j.biomaterials.2010.09.031 Cited on page(s) 60

[451] Kuhl, P.R. and L.G. Griffith-Cima, *Tethered epidermal growth factor as a paradigm for growth factor-induced stimulation from the solid phase.* Nat Med, 1996. **2**(9): p. 1022–7. DOI: 10.1038/nm0996-1022 Cited on page(s) 60

[452] Kapur, T.A. and M.S. Shoichet, *Chemically-bound nerve growth factor for neural tissue engineering applications.* J Biomater Sci Polym Ed, 2003. **14**(4): p. 383–94. DOI: 10.1163/156856203321478883 Cited on page(s)

[453] Fan, V.H., et al., *Tethered epidermal growth factor provides a survival advantage to mesenchymal stem cells.* Stem Cells, 2007. **25**(5): p. 1241–51. DOI: 10.1634/stemcells.2006-0320 Cited on page(s)

[454] Nakajima, M., et al., *Combinatorial protein display for the cell-based screening of biomaterials that direct neural stem cell differentiation.* Biomaterials, 2007. **28**(6): p. 1048–60. DOI: 10.1016/j.biomaterials.2006.10.004 Cited on page(s)

[455] Alberti, K., et al., *Functional immobilization of signaling proteins enables control of stem cell fate.* Nat Methods, 2008. **5**(7): p. 645–50. DOI: 10.1038/nmeth.1222 Cited on page(s) 60

[456] Aizawa, Y., et al., *The effect of immobilized platelet derived growth factor AA on neural stem/progenitor cell differentiation on cell-adhesive hydrogels.* Biomaterials, 2008. **29**(35): p. 4676–83. DOI: 10.1016/j.biomaterials.2008.08.018 Cited on page(s) 60

[457] Shen, Y.H., M.S. Shoichet, and M. Radisic, *Vascular endothelial growth factor immobilized in collagen scaffold promotes penetration and proliferation of endothelial cells.* Acta Biomater, 2008. **4**(3): p. 477–89. DOI: 10.1016/j.actbio.2007.12.011 Cited on page(s)

[458] Wall, S.T., et al., *Multivalency of Sonic hedgehog conjugated to linear polymer chains modulates protein potency.* Bioconjug Chem, 2008. **19**(4): p. 806–12. DOI: 10.1021/bc700265k Cited on page(s) 60

[459] Leipzig, N.D., et al., *Functional immobilization of interferon-gamma induces neuronal differentiation of neural stem cells.* J Biomed Mater Res A, 2010. **93**(2): p. 625–33. DOI: 10.1002/jbm.a.32573 Cited on page(s) 60

[460] Kang, C.E., E.J. Gemeinhart, and R.A. Gemeinhart, *Cellular alignment by grafted adhesion peptide surface density gradients.* J Biomed Mater Res A, 2004. **71**(3): p. 403–11. DOI: 10.1002/jbm.a.30137 Cited on page(s) 60

[461] Adams, D.N., et al., *Growth cones turn and migrate up an immobilized gradient of the laminin IKVAV peptide.* J Neurobiol, 2005. **62**(1): p. 134–47. DOI: 10.1002/neu.20075 Cited on page(s)

[462] Guarnieri, D., et al., *Covalently immobilized RGD gradient on PEG hydrogel scaffold influences cell migration parameters.* Acta Biomater, 2010. **6**(7): p. 2532–9. DOI: 10.1016/j.actbio.2009.12.050 Cited on page(s)

[463] Bhangale, S.M., et al., *Biologically active protein gradients via microstamping.* Advanced Materials, 2005. **17**(7): p. 809-+. DOI: 10.1002/adma.200400547 Cited on page(s) 60

[464] Yu, L.M., J.H. Wosnick, and M.S. Shoichet, *Miniaturized system of neurotrophin patterning for guided regeneration.* J Neurosci Methods, 2008. **171**(2): p. 253–63. DOI: 10.1016/j.jneumeth.2008.03.023 Cited on page(s) 60, 63

[465] Yu, L.M., F.D. Miller, and M.S. Shoichet, *The use of immobilized neurotrophins to support neuron survival and guide nerve fiber growth in compartmentalized chambers.* Biomaterials, 2010. **31**(27): p. 6987–99. DOI: 10.1016/j.biomaterials.2010.05.070 Cited on page(s) 60, 63

[466] Sorribas, H., C. Padeste, and L. Tiefenauer, *Photolithographic generation of protein micropatterns for neuron culture applications.* Biomaterials, 2002. **23**(3): p. 893–900. DOI: 10.1016/S0142-9612(01)00199-5 Cited on page(s) 63

[467] Luo, Y. and M.S. Shoichet, *A photolabile hydrogel for guided three-dimensional cell growth and migration.* Nat Mater, 2004. **3**(4): p. 249–53. DOI: 10.1038/nmat1092 Cited on page(s) 61, 63

[468] Hoffmann, J.C. and J.L. West, *Three-dimensional photolithographic patterning of multiple bioactive ligands in poly(ethylene glycol) hydrogels.* Soft Matter, 2010. **6**(20): p. 5056–5063. DOI: 10.1039/c0sm00140f Cited on page(s) 63

[469] Jeon, H., et al., *Chemical Patterning of Ultrathin Polymer Films by Direct-Write Multiphoton Lithography.* Journal of the American Chemical Society, 2011. **133**(16): p. 6138–6141. DOI: 10.1021/ja200313q Cited on page(s)

[470] Wylie, R.G., et al., *Spatially controlled simultaneous patterning of multiple growth factors in three-dimensional hydrogels.* Nature Materials, 2011. **10**(10): p. 799–806. DOI: 10.1038/nmat3101 Cited on page(s) 63

[471] Hashi, H., et al., *Angiogenic activity of a fusion protein of the cell-binding domain of fibronectin and basic fibroblast growth factor.* Cell Struct Funct, 1994. **19**(1): p. 37–47. DOI: 10.1247/csf.19.37 Cited on page(s) 63

[472] Andrades, J.A., et al., *Engineering, expression, and renaturation of a collagen-targeted human bFGF fusion protein.* Growth Factors, 2001. **18**(4): p. 261–75. DOI: 10.3109/08977190109029115 Cited on page(s) 63

[473] Nishi, N., et al., *Collagen-binding growth factors: Production and characterization of functional fusion proteins having a collagen-binding domain.* Proceedings of the National Academy of Sciences of the United States of America, 1998. **95**(12): p. 7018–7023. DOI: 10.1073/pnas.95.12.7018 Cited on page(s) 63

[474] Hayashi, M., M. Tomita, and K. Yoshizato, *Production of EGF-collagen chimeric protein which shows the mitogenic activity.* Biochimica Et Biophysica Acta-General Subjects, 2001. **1528**(2–3): p. 187–195. DOI: 10.1016/S0304-4165(01)00187-8 Cited on page(s) 63

[475] Lin, H., et al., *The effect of collagen-targeting platelet-derived growth factor on cellularization and vascularization of collagen scaffolds.* Biomaterials, 2006. **27**(33): p. 5708–5714. DOI: 10.1016/j.biomaterials.2006.07.023 Cited on page(s) 63

[476] Ota, T., et al., *A fusion protein of hepatocyte growth factor enhances reconstruction of myocardium in a cardiac patch derived from porcine urinary bladder matrix.* Journal of Thoracic and Cardiovascular Surgery, 2008. **136**(5): p. 1309–1317. DOI: 10.1016/j.jtcvs.2008.07.008 Cited on page(s) 63

[477] Sun, W., et al., *Promotion of peripheral nerve growth by collagen scaffolds loaded with collagen-targeting human nerve growth factor-beta.* J Biomed Mater Res A, 2007. **83**(4): p. 1054–61. DOI: 10.1002/jbm.a.31417 Cited on page(s) 63

[478] McCaig, C.D., *Nerve guidance - a role for bio-electric fields.* Progress in Neurobiology, 1988. **30**(6): p. 449–468. DOI: 10.1016/0301-0082(88)90031-7 Cited on page(s) 63

[479] Jaffe, L.F. and M.M. Poo, *Neurites grow faster towards the cathode than the anode in a steady field.* J Exp Zool, 1979. **209**(1): p. 115–28. DOI: 10.1002/jez.1402090114 Cited on page(s) 63

[480] Erskine, L. and C.D. McCaig, *Growth cone neurotransmitter receptor activation modulates electric field-guided nerve growth.* Dev Biol, 1995. **171**(2): p. 330–9. DOI: 10.1006/dbio.1995.1285 Cited on page(s)

[481] McCaig, C.D., *Nerve branching is induced and oriented by a small applied electric field.* J Cell Sci, 1990. **95 (Pt 4)**: p. 605–15. Cited on page(s)

[482] McCaig, C.D., L. Sangster, and R. Stewart, *Neurotrophins enhance electric field-directed growth cone guidance and directed nerve branching.* Dev Dyn, 2000. **217**(3): p. 299–308. DOI: 10.1002/(SICI)1097-0177(200003)217:3%3C299::AID-DVDY8%3E3.0.CO;2-G Cited on page(s)

[483] Patel, N. and M.M. Poo, *Orientation of neurite growth by extracellular electric fields.* J Neurosci, 1982. **2**(4): p. 483–96. Cited on page(s) 63

[484] Wan, L.D., R. Xia, and W.L. Ding, *Low-frequency electrical stimulation improves neurite outgrowth of dorsal root ganglion neurons in vitro via upregulating Ca(2+)-mediated brain-derived neurotrophic factor expression.* Neural Regeneration Research, 2010. **5**(16): p. 1256–1260. DOI: 10.3969/j.issn.1673-5374.2010.16.010 Cited on page(s) 63

[485] Wood, M. and R.K. Willits, *Short-duration, DC electrical stimulation increases chick embryo DRG neurite outgrowth.* Bioelectromagnetics, 2006. **27**(4): p. 328–331. DOI: 10.1002/bem.20214 Cited on page(s)

[486] Wood, M.D. and R.K. Willits, *Applied electric field enhances DRG neurite growth: influence of stimulation media, surface coating and growth supplements.* Journal of Neural Engineering, 2009. **6**(4). DOI: 10.1088/1741-2560/6/4/046003 Cited on page(s) 63

[487] Rajnicek, A.M., K.R. Robinson, and C.D. McCaig, *The direction of neurite growth in a weak DC electric field depends on the substratum: Contributions of adhesivity and net surface charge.* Developmental Biology, 1998. **203**(2): p. 412–423. DOI: 10.1006/dbio.1998.9039 Cited on page(s) 64

[488] Ariza, C.A., et al., *The Influence of Electric Fields on Hippocampal Neural Progenitor Cells.* Stem Cell Reviews and Reports, 2010. **6**(4): p. 585–600. DOI: 10.1007/s12015-010-9171-0 Cited on page(s) 64

[489] Arocena, M., et al., *A Time-Lapse and Quantitative Modelling Analysis of Neural Stem Cell Motion in the Absence of Directional Cues and in Electric Fields.* Journal of Neuroscience Research, 2010. **88**(15): p. 3267–3274. DOI: 10.1002/jnr.22502 Cited on page(s) 64

[490] Al-Majed, A.A., et al., *Brief electrical stimulation promotes the speed and accuracy of motor axonal regeneration.* J Neurosci, 2000. **20**(7): p. 2602–8. Cited on page(s) 64

[491] Sisken, B.F., et al., *Stimulation of rat sciatic nerve regeneration with pulsed electromagnetic fields.* Brain Res, 1989. **485**(2): p. 309–16. DOI: 10.1016/0006-8993(89)90575-1 Cited on page(s) 64

[492] Langer, R. and J.P. Vacanti, *Tissue engineering.* Science, 1993. **260**(5110): p. 920–926. DOI: 10.1126/science.8493529 Cited on page(s) 65

[493] Schmidt, C.E. and J.B. Leach, *Neural tissue engineering: Strategies for repair and regeneration.* Annual Review of Biomedical Engineering, 2003. **5**: p. 293–347. DOI: 10.1146/annurev.bioeng.5.011303.120731 Cited on page(s) 4

[494] Stern, C.D., *Neural induction: old problem, new findings, yet more questions.* Development, 2005. **132**(9): p. 2007–2021. DOI: 10.1242/dev.01794 Cited on page(s) 35

[495] Jones, D.L. and A.J. Wagers, *No place like home: anatomy and function of the stem cell niche.* Nature Reviews Molecular Cell Biology, 2008. **9**(1): p. 11–21. DOI: 10.1038/nrm2319 Cited on page(s) 40

[496] Lim, D.A., et al., *Noggin antagonizes BMP signaling to create a niche for adult neurogenesis.* Neuron, 2000. **28**(3): p. 713–726. DOI: 10.1016/S0896-6273(00)00148-3 Cited on page(s)

[497] Walker, M.R., K.K. Patel, and T.S. Stappenbeck, *The stem cell niche.* Journal of Pathology, 2009. **217**(2): p. 169–180. DOI: 10.1002/path.2474 Cited on page(s) 40

Authors' Biographies

ASHLEY E. WILKINSON

Ashley E. Wilkinson is currently pursuing her PhD in Chemical and Biomolecular Engineering at the University of Akron (Akron, Ohio). She received her Bachelor of Science in Biomedical Engineering in 2010 from the University of Akron with a specialization in drug delivery and tissue engineering. During her undergraduate studies, Ashley participated in cooperative education at DePuy Orthopaedics as a product development engineer. Her current research interests include neuroregenerative strategies in the brain and spinal cord, specifically stem cell differentiation via specific chemical and mechanical stimulation.

ALEESHA M. MCCORMICK

Aleesha M. McCormick obtained her Bachelor of Science at Kent State University (Kent, Ohio) in Integrated Science Education and taught high school chemistry and physics for two years. She is currently pursuing a PhD at the University of Akron (Akron, OH) in Chemical and Biomolecular Engineering. Her research area focuses on recombinant protein analysis and production for utilization in axon guidance and neural regenerative applications as well as examining potential cell sources for neuronal differentiation.

NIC D. LEIPZIG

Nic D. Leipzig is the Iredell Chair Assistant Professor in Chemical and Biomolecular Engineering at the University of Akron (Akron, OH). He received a Bachelor of Engineering in Chemical Engineering from McGill University (Montreal, Quebec) in 2001 and a PhD in Bioengineering from Rice University (Houston, Texas) in 2006. During his PhD he studied the biomechanics of single chondrocytes, or cartilage cells, explored how growth factors change both the cytoskeleton and the material properties of chondrocytes, developed new methods for measuring gene expression in single cells and utilized these techniques to be the first to successfully demonstrate gene expression changes by mechanotransduction in single chondrocytes. He was a postdoctoral fellow at the University of Toronto (Toronto, Ontario) in the department of Chemical Engineering and Applied Chemistry from 2006 to 2009, where he developed hydrogel systems to enable precise control of the cell microenvironment, or niche, in order to guide the differentiation of adult stem cells. He has also revealed that substrate stiffness can influence neural stem cell proliferation and differentiation and demonstrated the advantages of covalently attaching growth factors for precisely guiding stem cell differentiation. Dr. Leipzig's current research is pioneering approaches for tissue engineering of the central nervous system utilizing engineered biomaterials, incorporating niche level stimuli and new stem cell sources.

Printed in the United States
by Baker & Taylor Publisher Services